DYNAMICS STUDY PACK

CHAPTER REVIEWS, FREE-BODY DIAGRAM WORKBOOK, COMPANION WEBSITE

D0103077

ENGINEERING MECHANICS

DYNAMICS

TWELFTH EDITION

R. C. HIBBELER

PRENTICE HALL
Upper Saddle River, NJ 07458

Editorial Director of Computer Science and Engineering: *Marcia J. Horton*
Acquisitions Editor: *Tacy Quinn*
Associate Editor: *Dee Bernhard*
Editorial Assistant: *Bernadette Marciniak*
Director of Marketing: *Margaret Waples*
Marketing Manager: *Tim Galligan*
Senior Managing Editor: *Scott Disanno*
Production Editor: *Irwin Zucker*
Art Director: *Kenny Beck*
Art Editor: *Greg Dulles*
Media Editor: *Daniel Sandin*
Manufacturing Manager: *Alan Fischer*
Manufacturing Buyer: *Lisa McDowell*

Pearson Education Ltd., *London*
Pearson Education Australia Pty. Ltd., *Sydney*
Pearson Education Singapore, Pte. Ltd.
Pearson Education North Asia Ltd., *Hong Kong*
Pearson Education Canada, Inc., *Toronto*
Pearson Educación de Mexico, S.A. de C.V.
Pearson Education—Japan, *Tokyo*
Pearson Education Malaysia, Pte. Ltd.
Pearson Education, *Upper Saddle River, New Jersey*

Prentice Hall
is an imprint of

www.pearsonhighered.com

10 9 8 7 6 5 4 3 2 1
ISBN-13: 978-0-13-609195-0
ISBN-10:   0-13-609195-4

Contents

What's in This Package

The Dynamics Study Pack was designed to help students improve their study skills. It consists of three study components—a chapter-by-chapter review, a free-body diagram workbook, and an access code for the Companion Website.

- **Chapter-by-Chapter Review and Free-Body Diagram Workbook**—Prepared by Peter Schiavone of the University of Alberta. This resource contains chapter-by-chapter *Dynamics* review, including key points, equations, and check up questions. The Free-Body Diagram Workbook steps students through numerous free-body diagram problems that include full explanations and solutions.

- **Companion Website**—Located at *www.prenhall.com/hibbeler*, the Companion Website includes the following resources:

  - **Video Solutions** — Complete, step-by-step solution walkthroughs of representative homework problems from each section in the Dynamics text. Developed by Professor Edward Berger of University of Virginia, Video Solutions offer:

    - **Fully-worked Solutions** — Showing every step of representative homework problems, to help students make vital connections between concepts.

    - **Self-paced Instruction** — Students can navigate each problem and select, play, rewind, fast-forward, stop, and jump-to sections within each problem's solution.

    - **24/7 Access** — Help whenever students need it with over 20 hours of helpful review.

  - **1000 Supplemental *Statics* and *Dynamics* Problems** — A self-study resource, these supplemental problems have complete solutions provided.

  - **MATLAB and Mathcad Tutorials** — Focused on using MATLAB and Mathcad in Engineering Mechanics, these tutorials are keyed to the text.

To log in to the Companion Website, follow the instructions on the access card included with this study pack.

Preface

This supplement is divided into two parts. Part I provides a section-by-section, chapter-by-chapter summary of the key concepts, principles and equations from R. C. Hibbeler's text, *Engineering Mechanics–Dynamics*, Twelfth Edition. Part II is a workbook which explains how to draw and use free-body diagrams when solving problems in *Dynamics*.

Part I: Chapter-by-Chapter Summaries

This part of the supplement provides a section-by-section, chapter-by-chapter summary of the key concepts, principles and equations from R. C. Hibbeler's text, *Engineering Mechanics–Dynamics*, Twelfth Edition. We follow the same section and chapter order as that used in the text and summarize important concepts from each section in easy-to-understand language. We end each chapter summary with a simple set of review questions designed to see if the student has understood the key concepts and chapter objectives.

This section of the supplement will be useful both as a quick reference guide for important concepts and equations when solving problems in, for example, homework assignments or laboratories and also as a handy review when preparing for any quiz, test, or examination.

Part II: Free-Body Diagram Workbook

A thorough understanding of how to draw and use a free-body diagram is absolutely essential when solving problems in mechanics.

This workbook consists mainly of a collection of problems intended to give the student practice in drawing and using free-body diagrams when solving problems in *Dynamics*.

All the problems are presented as *tutorial* problems with the solution only partially complete. The student is then expected to complete the solution by "filling in the blanks" in the spaces provided. This gives the student the opportunity to *build free-body diagrams in stages* and extract the relevant information from them when formulating equilibrium equations. Earlier problems provide students with partially drawn free-body diagrams and lots of hints to complete the solution. Later problems are more advanced and are designed to challenge the student more. The complete solution to each problem can be found on the back of the page. The problems are chosen from two-dimensional theories of particle and rigid body mechanics. Once the ideas and concepts developed in these problems have been understood and practiced, the student will find that they can be extended in a relatively straightforward manner to accommodate the corresponding three-dimensional theories.

The workbook begins with a brief primer on free-body diagrams: where they fit into the general procedure of solving problems in mechanics and why they are so important. Next follows a few examples to illustrate ideas and then the workbook problems.

For best results, the student should read the primer and then, beginning with the simpler problems, try to complete and understand the solution to each of the subsequent problems. The student should avoid the temptation to immediately look at the completed solution on the back of the page. This solution should be accessed only as a last resort (after the student has struggled to the point of giving up), or to check the student's own solution after the fact. The idea behind this is very simple: *we learn most when we* **do** *the thing we are trying to learn*—reading through someone else's solution is not the same as actually working through the problem. In the former, the student gains *information*, in the latter the student gains *knowledge*. For example, how many people learn to swim or drive a car by reading an instruction manual?

Consequently, since the workbook is based on **doing**, the student who persistently solves the problems in the workbook will ultimately gain a thorough, usable knowledge of how to draw and use free-body diagrams.

PETER SCHIAVONE

PART I

Section-By-Section, Chapter-By-Chapter

Summaries with Review Questions and

Answers

12

Kinematics of a Particle

MAIN GOALS OF THIS CHAPTER:

- To introduce the concepts of position, displacement, velocity, and acceleration.
- To study particle motion along a straight line and represent this motion graphically.
- To investigate particle motion along a curved path using different coordinate systems.
- To present an analysis of dependent motion of two particles.
- To examine the principles of relative motion of two particles using translating axes.

12.1 INTRODUCTION

Mechanics is that branch of the physical sciences concerned with the behavior of bodies subjected to the action of forces. The subject of mechanics is divided into two parts:

- *statics* - the study of objects in equilibrium (objects either at rest or moving with a constant velocity).
- *dynamics* - the study of objects with accelerated motion.

  ◆ The subject of *dynamics* is often itself divided into two parts:

    – *kinematics* - treats only the *geometric aspects* of the motion.
    – *kinetics* - analysis of *forces* causing the motion.

- In this chapter, we study the *kinematics of a particle* - recall that a particle has *a mass but negligible size and shape*. Therefore, we limit discussion to those objects that have dimensions that are of no consequence in the analysis of the motion. Such objects may be considered as particles, provided motion of the body is characterized by motion of its mass center and any rotation of the body is neglected.

12.2 RECTILINEAR KINEMATICS: CONTINUOUS MOTION

- *Rectilinear Kinematics* refers to straight line motion. The kinematics of a particle is characterized by specifying, at any given instant, the particle's *position, velocity and acceleration.*

    - **Position.** The *position* of the particle is represented by an algebraic *scalar s* (*the position coordinate*).
    - **Displacement.** The *displacement* of the particle is a *vector* $\triangle \mathbf{r}$ defined as the change in the particle's position vector $\mathbf{r}$.
    - **Velocity.** The *velocity* of the particle is a *vector*.

        * The *average velocity* is the displacement divided by time i.e., $\mathbf{v}_{avg} = \frac{\triangle \mathbf{r}}{\triangle t}$.

        * The *instantaneous velocity* is $\mathbf{v} = \frac{d\mathbf{r}}{dt}$.

        * *Speed* refers to the *magnitude* of velocity. and is written as $v = |\mathbf{v}| = \frac{ds}{dt}$.

        * *Average speed* is the total distance divided by the total time (different from average velocity which is displacement divided by time).

    - **Acceleration.** The *acceleration* of the particle is a *vector* $\mathbf{a} = \frac{d\mathbf{v}}{dt}$. It's magnitude is written as $a = \frac{dv}{dt} = \frac{d^2 s}{dt^2}$.

        * In rectilinear kinematics, the acceleration is negative when the particle is slowing down or decelerating.
        * A particle can have an acceleration and yet have zero velocity.
        * The relationship $a\,ds = v\,dv$ is derived from $a = \frac{dv}{dt}$ and $v = \frac{ds}{dt}$ by eliminating dt.

CONSTANT ACCELERATION

- Let $a = a_c = $ constant. Assume that $v = v_0$ and $s = s_0$ at time $t = 0$. Then

$$v = v_0 + a_c t \text{ (speed as a function of time),} \tag{12.0}$$

$$s = s_0 + v_0 t + \frac{1}{2} a_c t^2 \text{ (position as a function of time),} \tag{12.1}$$

$$v^2 = v_0^2 + 2a_c (s - s_0) \text{ (speed as a function of position).} \tag{12.2}$$

PROCEDURE FOR SOLVING PROBLEMS

The equations of rectilinear kinematics should be applied as follows:

- **Coordinate System**

    - Establish a position coordinate s along the path and specify its *fixed* origin and positive direction.
    - Since motion is along a straight line, the particle's position, velocity and acceleration can be represented as algebraic scalars. For analytical work, the sense of s, v and a is then determined from their algebraic signs.
    - The positive sense for each scalar can be indicated by an arrow shown alongside each kinematic equation as it is applied.

- **Kinematic Equations.**

    - If a relationship is known between any two of the four variables a, v, s and t, then a third variable can be obtained by using one of the kinematic equations $a = \frac{dv}{dt}$, $v = \frac{ds}{dt}$ or $a\,ds = v\,dv$, which relates all three variables.
    - Whenever integration is performed, it is important that the position and velocity be known at a given instant in order to evaluate either the constant of integration if an indefinite integral is used, or the limits of integration if a definite integral is used.
    - **Note** that Eqs. (12.0)–(12.2) apply **only** *when the acceleration is constant.*

12.3 RECTILINEAR KINEMATICS: ERRATIC MOTION

When a particle's motion during a time period is erratic, it may be difficult to obtain a continuous function to describe its position, velocity or acceleration. Instead, the motion may best be described graphically using a series of curves that can be generated experimentally by computer. There are several frequently occurring situations:

- **Given** $s - t$ **Graph, Construct** $v - t$ **Graph.** If the position of a particle can be plotted over time ($s - t$ graph), the particle's velocity as a function of time ($v - t$ graph) can be obtained by measuring the slope of the $s - t$ graph.

$$\frac{ds}{dt} = v$$
$$\text{slope of } s - t \text{ graph } = velocity$$

- **Given** $v - t$ **Graph, Construct** $a - t$ **Graph.** When the particle's $v - t$ graph is known, the particle's acceleration as a function of time ($a - t$ graph) can be obtained by measuring the slope of the $v - t$ graph.

$$\frac{dv}{dt} = a$$
$$\text{slope of } v - t \text{ graph } = acceleration$$

- **Given** $a - t$ **Graph, Construct** $v - t$ **Graph.** When the particle's $a - t$ graph is given, the $v - t$ graph may be constructed by:

$$\triangle v = \int a\, dt$$
$$\text{change in velocity } = \text{ area under } a - t \text{ graph}$$

- **Given** $v - t$ **Graph, Construct** $s - t$ **Graph.** When the particle's $v - t$ graph is given, the $s - t$ graph may be constructed by:

$$\triangle s = \int v\, dt$$
$$\text{displacement } = \text{ area under } v - t \text{ graph}$$

- **Given** $a - s$ **Graph, Construct** $v - s$ **Graph.** When the particle's $a - s$ graph can be constructed, the $v - s$ graph may be obtained:

$$\frac{1}{2}\left(v_1^2 - v_0^2\right) = \int_{s_0}^{s_1} a\, ds$$
$$= \text{ area under } a - s \text{ graph}$$

- **Given** $v - s$ **Graph, Construct** $a - s$ **Graph.** When the particle's $v - s$ graph is known the acceleration at any position s can be obtained:

$$a = v\frac{dv}{ds}$$
$$\text{acceleration } = \text{ velocity } \times \text{ slope of } v - s \text{ graph}$$

12.4 GENERAL CURVILINEAR MOTION

Curvilinear motion occurs when the particle moves along a *curved path*.

- **Position.** The *position* of the particle is described by the position *vector* $\mathbf{r}(t)$. This vector is a function of time since *both its magnitude and direction* change as the particle moves along its path (described by the path function $s(t)$).

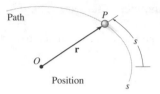

- **Velocity**. The *velocity* of the particle is described by the *vector* $\mathbf{v}(t) = \dfrac{d\mathbf{r}}{dt}$.

    - **Speed.** The *speed* v is the magnitude of $\mathbf{v}$ and can be obtained by differentiating the path function $s(t)$ with respect to time i.e., $v = \dfrac{ds}{dt}$.

- **Acceleration**. The *acceleration* of the particle is described by the *vector* $\mathbf{a}(t) = \dfrac{d\mathbf{v}}{dt} = \dfrac{d^2\mathbf{r}}{dt^2}$.

IMPORTANT POINTS

- Curvilinear motion can cause changes in *both* the magnitude and direction of the position, velocity and acceleration vectors.
- The velocity vector is *always directed tangent* to the path.
- In general, the acceleration vector is *not* tangent to the path, but rather, it is tangent to the hodograph.

12.5 CURVILINEAR MOTION: RECTANGULAR COMPONENTS

Using a fixed $xyz-$ frame of reference:

- **Position**.

$$\mathbf{r}(t) = x(t)\mathbf{i} + y(t)\mathbf{j} + z(t)\mathbf{k},$$
$$\text{Magnitude} = r = \sqrt{x^2 + y^2 + z^2},$$
$$\text{Direction given by} = \frac{\mathbf{r}}{r}.$$

- **Velocity**.

$$\mathbf{v}(t) = \frac{d\mathbf{r}}{dt} = \frac{d}{dt}[x(t)\mathbf{i}] + \frac{d}{dt}[y(t)\mathbf{j}] + \frac{d}{dt}[z(t)\mathbf{k}],$$
$$= \frac{d}{dt}[x(t)]\mathbf{i} + \frac{d}{dt}[y(t)]\mathbf{j} + \frac{d}{dt}[z(t)]\mathbf{k}$$
$$= v_x\mathbf{i} + v_y\mathbf{j} + v_z\mathbf{k}$$
$$\text{Magnitude} = v = \sqrt{v_x^2 + v_y^2 + v_z^2}, \ \text{Direction:} \ \frac{\mathbf{v}}{v}, \text{always tangent to path.}$$

- **Acceleration**.

$$\mathbf{a}(t) = \frac{d\mathbf{v}}{dt} = a_x\mathbf{i} + a_y\mathbf{j} + a_z\mathbf{k}$$
$$\text{where } a_x = \dot{v}_x = \ddot{x}, \ a_y = \dot{v}_y = \ddot{y}, \ a_z = \dot{v}_z = \ddot{z},$$
$$\text{Magnitude} = a = \sqrt{a_x^2 + a_y^2 + a_z^2}, \ \text{Direction:} \ \frac{\mathbf{a}}{a}.$$

12.6 MOTION OF A PROJECTILE

The free-flight motion of a projectile is often studied in terms of its rectangular components, since the projectile's acceleration *always* acts in the vertical direction.

PROCEDURE FOR SOLVING PROBLEMS

- Coordinate System

  - Establish the fixed x, y coordinate axes and sketch the trajectory of the particle. Between any two points on the path specify the problem data and the three unknowns. In all cases the acceleration of gravity acts downward. The particle's initial and final velocities should be represented in terms of their x and y components.
  - Remember that positive and negative position, velocity and acceleration components always act in accordance with their associated coordinate directions.

- Kinematic Equations

  - Depending upon the known data and what is to be determined, a choice should be made as to which three of the following four equations should be applied between the two points on the path to obtain the most direct solution to the problem.

- Horizontal Motion

  - The velocity in the horizontal or x direction is *constant* i.e., $(v_x) = (v_0)_x$ and

  $$x = x_0 + (v_0)_x\, t.$$

- Vertical Motion

  - In the vertical or $y-$direction *only two* of the following three equations can be used for solution:

  $$v_y = (v_0)_y - a_c t,$$
  $$y = y_0 + (v_0)_y t + \frac{1}{2}a_c t^2,$$
  $$v_y^2 = (v_0)_y^2 + 2a_c\, (y - y_0).$$

  - For example, if the particle's final velocity v_y is not needed, then the first and third of these equations (for y) will not be useful.

12.7 CURVILINEAR MOTION: NORMAL AND TANGENTIAL COMPONENTS

IMPORTANT POINTS

Coordinate System

- Provided the *path* of the particle is *known*, we establish a set of n and t coordinates having a *fixed origin* which is coincident with the particle at the instant considered.
- The positive tangent axis always acts in the direction of motion and the positive normal axis is directed towards the path's center of curvature.
- The n and t axes are particularly advantageous for studying the velocity and acceleration of the particle, because the velocity $\mathbf{v}$ and the acceleration $\mathbf{a}$ are expressed by the equations

$$\mathbf{v} = v\mathbf{u}_t, \quad v = \dot{s}, \tag{12.3}$$
$$\mathbf{a} = a_t\mathbf{u}_t + a_n\mathbf{u}_n, \tag{12.4}$$

$$a_t = \dot{v} \quad \text{or} \quad a_t ds = v dv, \tag{12.5}$$

$$a_n = \frac{v^2}{\rho}. \tag{12.6}$$

Velocity

- The particle's *velocity* is always tangent to the path.
- The magnitude of velocity (speed) is found from the time derivative of the path function $v = \dot{s}$.

Tangential Acceleration

- The tangential component of acceleration is the result of the time rate of change in the magnitude of velocity. This component acts in the positive s−direction if the particle's speed is increasing or in the opposite direction if the speed is decreasing.
- The relations between a_t v and s are the same as for rectilinear motion:

$$a_t = \dot{v}, \quad a_t ds = v dv.$$

- **If a_t is constant**, $a_t = (a_t)_c$, the above equations, when integrated, yield

$$s = s_0 + v_0 t + \frac{1}{2}(a_t)_c t^2$$
$$v = v_0 + (a_t)_c t$$
$$v^2 = v_0^2 + 2(a_t)_c(s - s_0)$$

Normal Acceleration

- The normal component of acceleration is the result of the time rate of change in the direction of the particle's velocity. This component is always directed toward the center of curvature of the path i.e., along the positive n axis.
- The magnitude of this component is determined from $a_n = \frac{v^2}{\rho}$.
- If the path is expressed as $y = f(x)$, the radius of curvature ρ at any point on the path is determined from the equation

$$\rho = \frac{[1 + (dy/dx)^2]^{\frac{3}{2}}}{\left|\dfrac{d^2 y}{dx^2}\right|}.$$

12.8 CURVILINEAR MOTION: CYLINDRICAL COMPONENTS

In some problems it is often convenient to express the path of motion in terms of cylindrical coordinates r, θ, z. If motion is restricted to the plane, polar coordinates r and θ are used.

IMPORTANT POINTS

Coordinate System

- Polar coordinates are particularly suitable for solving problems for which data regarding the angular motion of the radial coordinate r is given to describe the particle's motion.
- To use polar coordinates, the origin is established at a fixed point, and the radial line r is directed to the particle.
- The transverse coordinate θ is measured counterclockwise from a fixed reference line to the radial line.

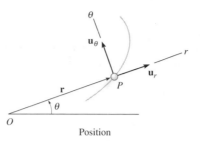

Position

Position, Velocity and Acceleration

- The position of the particle is defined by the position vector $\mathbf{r} = r\mathbf{u}_r$.
- The velocity and acceleration of the particle are given, respectively, by

$$\mathbf{v} = v_r\mathbf{u}_r + v_\theta\mathbf{u}_\theta$$
$$\mathbf{a} = a_r\mathbf{u}_r + a_\theta\mathbf{u}_\theta$$

- Once r and the four time derivatives $\dot{r}, \ddot{r}, \dot{\theta}$, and $\ddot{\theta}$ have been evaluated at the instant considered, their values can be substituted into the following equations to obtain the radial and transverse components of $\mathbf{v}$ and $\mathbf{a}$.

$$v_r = \dot{r}, \quad v_\theta - r\dot{\theta}, \tag{12.7}$$
$$a_r = \ddot{r} - r\dot{\theta}^2, \quad a_\theta = r\ddot{\theta} + 2\dot{r}\dot{\theta}. \tag{12.8}$$

- If it is necessary to take the time derivatives of $r = f(\theta)$, it is very important to use the chain rule.
- Motion in three-dimensions requires a simple extension of the above formula to

$$\text{Position: } \mathbf{r} = r\mathbf{u}_r + z\mathbf{u}_z,$$
$$\text{Velocity: } \mathbf{v} = \dot{r}\mathbf{u}_r + r\dot{\theta}\mathbf{u}_\theta + \dot{z}\mathbf{u}_z,$$
$$\text{Acceleration: } a = (\ddot{r} - r\dot{\theta}^2)\mathbf{u}_r + (r\ddot{\theta} + 2\dot{r}\dot{\theta})\mathbf{u}_\theta + \ddot{z}\mathbf{u}_z.$$

- **Note**: If the particle travels in a *circular* path, $r = $ constant so that $\dot{r} = \ddot{r} = 0$ and the formulas (12.7)–(12.8) simplify considerably.

12.9 ABSOLUTE DEPENDENT MOTION ANALYSIS OF TWO PARTICLES

In some problems, the motion of one particle will *depend* on the corresponding motion of another particle. This dependency commonly occurs if the particles are interconnected by inextensible cords which are wrapped around pulleys. When each particle moves along a *rectilinear path*, itudes of the velocity and acceleration of the particles will change, not their line of direction. The following procedure can be used:

PROCEDURE FOR SOLVING PROBLEMS

- **Position-Coordinate Equation**

  - Establish position coordinates which have their origin located at a *fixed* point or datum.
  - The coordinates are directed along the path of motion and extend to a point having the same motion as each of the particles.
  - It is *not necessary* that the *origin* be the same for each of the coordinates; however, it is *important* that each coordinate axis selected be directed along the *path of motion* of the particle.
  - Using geometry or trigonometry, relate the coordinates to the total length of the cord, l_T or to that portion of cord, l, which *excludes* the segments that do not change length as the particles move - such as arc segments

wrapped over pulleys.

- If a problem involves a *system* of two or more cords wrapped around pulleys, then the position of a point on one cord must be related to the position of a point on another cord. Separate equations are written for a fixed length of each cord of the system and the positions of the two particles are then related by these equations.

- **Time Derivatives**

  - Two successive time derivatives of the position-coordinate equations yield the required velocity and acceleration equations which relate the motions of the particles.

  - The signs of the terms in these equations will be consistent with those that specify the positive and negative sense of the position coordinates.

12.10 RELATIVE-MOTION OF TWO PARTICLES USING TRANSLATING AXES

When the path of motion for a particle is complicated, it may be feasible to analyze the motion in parts by using two or more frames of reference.

PROCEDURE FOR SOLVING PROBLEMS

- When applying the relative-position equation, $\mathbf{r}_B = \mathbf{r}_A + \mathbf{r}_{B/A}$, it is first necessary to specify the locations of the fixed x, y, z and translating x', y', z' axes.

- Usually, the origin A of the translating axes is located at a point having a *known position*, $\mathbf{r}_A$.

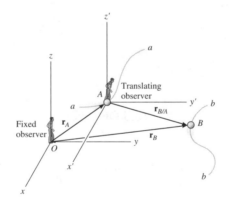

- A graphical representation of the vector addition $\mathbf{r}_B = \mathbf{r}_A + \mathbf{r}_{B/A}$ can be shown, and both the known and unknown quantities labeled on this sketch.

- Since vector addition forms a triangle, there can be at most *two unknowns*, represented by the magnitudes and/or directions of the vector quantities.

- These unknowns can be solved for either graphically, using trigonometry (sine, cosine law), or by resolving each of the three vectors $\mathbf{r}_B$, $\mathbf{r}_A$ and $\mathbf{r}_{B/A}$ into rectangular or Cartesian components, thereby generating a set of scalar equations.

- The relative-motion equations $\mathbf{v}_B = \mathbf{v}_A + \mathbf{v}_{B/A}$ and $\mathbf{a}_B = \mathbf{a}_A + \mathbf{a}_{B/A}$ are applied in the same manner as mentioned above except that in this case the origin O of the fixed x, y, z axes does not have to be specified.

HELPFUL TIPS AND SUGGESTIONS

- The most effective way to learn the principles of dynamics is to *solve problems* - in a logical and orderly manner (*practice is the key!*).
- *Remember* that in solving problems from engineering mechanics you are solving real practical problems and producing real data with physical significance. Thus, you are responsible for making sure your results are *correct*, *consistent* and *well-presented*. Get into the habit of doing this *now* so that it will become second nature by the time you graduate. In the world of professional engineering you have a *responsibility* to your profession and to the many people that will use the product you will help to design, manufacture or implement.

REVIEW QUESTIONS

1. Can the kinematics of a particle can be regarded as the same as the kinematics of a point?
2. Is the velocity of a point always tangent to its path?
3. Is the acceleration of a point always tangent to its path?
4. If a ball is travelling in a horizontal circle at constant speed, does the center of the ball have zero acceleration?
5. Is the velocity of a point independent of the reference frame chosen to express the position of the point?
6. When does the acceleration of a particle always have a zero normal component?
7. If a particle in rectilinear motion has zero speed at some instant in time, is the acceleration necessarily zero at the same instant?
8. Can a particle have $\ddot{r} = 0$ but still have a nonvanishing radial component of acceleration?

13

Kinetics of a Particle:
Force and Acceleration

MAIN GOALS OF THIS CHAPTER:

- To state Newton's Laws of Motion and Gravitational attraction and to define mass and weight.
- To analyze the accelerated motion of a particle using the equation of motion with different coordinate systems.
- To investigate central-force motion and apply it to problems in space mechanics.

13.1 NEWTON'S SECOND LAW OF MOTION

- *Newton's second law of motion* states that the *unbalanced force* on a particle causes it to accelerate. If the mass of the particle is m and its velocity is $\mathbf{v}$, the second law can be written as:

$$\mathbf{F} = \frac{d}{dt}(m\mathbf{v}) = m\mathbf{a}. \tag{13.0}$$

Equation (13.0) is referred to as the *equation of motion* and is one of the most important formulations in mechanics. It's validity is based *solely* on experimental evidence.

- **Newton's Law of Gravitational Attraction:** The mutual attraction between any two particles is given by

$$F = G\frac{m_1 m_2}{r^2}$$

where F is the force of attraction between the two particles, G is the universal constant of gravitation ($G = 66.73(10^{-12})m^3/(kg \cdot s^2)$), m_1, m_2 is the mass of each of the two particles and r is the distance between the centers of the two particles.

- **Mass and Weight.**

   Mass is a property of matter that provides a quantitative measure of its resistance to a change in velocity (see Equation(13.0)). Mass is an absolute quantity.

   Weight is a force that is caused by the earth's gravitation. It is not absolute; rather it depends on the altitude of the mass from the earth's surface.

The relationship between the weight W and mass m of a particle is given by $W = mg$ where g represents the *acceleration due to gravity.*

- **SI System of Units.** In the SI system, the mass of a body is specified in *kilograms (kg)* and the weight in *Newtons (N)* i.e.,

$$W = mg \ (N) \quad (g = 9.81 \ m/s^2).$$

- **FPS System of Units.** In the FPS system, the weight is specified in *pounds (lb)* and the mass in slugs *(slug)*. i.e.,

$$m = \frac{W}{g} \ (slug) \quad (g = 32.2 \ ft/s).$$

13.2 THE EQUATION OF MOTION

- When more than one force acts on a particle, the resultant force is determined by a vector summation of all the forces i.e., $\mathbf{F}_R = \sum \mathbf{F}$. For this more general case, the *equation of motion* (13.0) may be written as

$$\sum \mathbf{F} = m\mathbf{a}. \tag{13.1}$$

The magnitude and direction of each force acting on the particle (left-hand side of Equation (13.1) are identified using a *free-body diagram*. A kinetic diagram identifies the magnitude and direction of the vector $m\mathbf{a}$ (right-hand side of Equation (13.1)).

- **Inertial Frame.** Whenever the equation of motion (13.0) or (13.1) is applied, it is required that measurements of the acceleration be made from a *Newtonian or inertial frame of reference*.

A Newtonian or inertial frame of reference does not rotate and is either fixed or translates in a given direction with a constant velocity (zero acceleration).

13.3 EQUATION OF MOTION FOR A SYSTEM OF PARTICLES

- The sum of the external forces acting on a system of particles is equal to the total mass m of the particles times the acceleration of its center of mass G i.e.,

$$\sum \mathbf{F} = m\mathbf{a}_G. \tag{13.2}$$

SUMMARY

The equation of motion (13.0)–(13.2) is based on experimental evidence and is valid only when applied from an inertial frame of reference.

13.4 EQUATIONS OF MOTION: RECTANGULAR COORDINATES

- When a particle is moving relative to an inertial x, y, z frame of reference, the (vector) equation of motion (13.1) is equivalent to the following three *scalar* equations:

$$\sum F_x = ma_x,$$
$$\sum F_y = ma_y, \tag{13.3}$$
$$\sum F_z = ma_z.$$

Only the first two of these equations are used to specify the motion of a particle *constrained to move only in the* $x - y$ *plane.*

SOLVING PROBLEMS USING THE EQUATIONS OF MOTION (13.3)

- **Free-Body Diagram**

    - Select the inertial coordinate system. Rectangular or x, y, z coordinates are used to analyze problems involving *rectilinear motion*.

    - Draw the particle's *free-body diagram*. This makes it possible to resolve all the forces acting on the particle into their x, y, z components (for use in Equations (13.3)).

    - The direction and sense of the particle's acceleration **a** should also be established. If the senses of its components are unknown, assume they are in the *same direction* as the *positive* inertial coordinate axes.

    - The acceleration may be represented as the m**a** vector on the kinetic diagram.

    - Identify the unknowns in the problem.

- **Equations of Motion**

    - If the forces can be resolved directly from the free-body diagram, apply Equations (13.3).

    - If the geometry of the problem appears complicated, which often occurs in three dimensions, Cartesian vector analysis can be used for the solution.

    - **Friction**. If the particle contacts a rough surface, it may be necessary to use the frictional equation $F_f = \mu_k N$. Remember that F_f always acts to oppose the motion of the particle *relative to the surface it contacts*.

    - **Spring**. If the particle is connected to an elastic spring having negligible mass, the spring force F_s can be related to the deformation of the spring by the equation $F_s = ks$.

- **Kinematics**

    - If the velocity or position of the particle is to be found, it will be necessary to apply the proper kinematic equations once the particle's acceleration is determined from Equation (13.3).

    - If acceleration is a function of time, use $a = \dfrac{dv}{dt}$ and $v = \dfrac{ds}{dt}$, which, when integrated, yield the particle's speed and position.

    - If acceleration is a function of displacement, integrate $ads = vdv$ to obtain the speed as a function of position.

    - If acceleration is constant, use $v = v_0 + a_c t$, $s = s_0 + v_0 t + \frac{1}{2}a_c t^2$, $v^2 = v_0^2 + 2a_c (s - s_0)$ to determine the speed or position of the particle.

    - If the problem involves the dependent motion of several particles, use the procedure of Section 12.9 to relate their accelerations.

    - In all cases, make sure the positive inertial coordinate directions used for writing the kinematic equations are the *same* as those used for writing the equations of motion; otherwise, simultaneous solution of the equations will result in errors.

    - If the solution for an unknown vector component yields a negative scalar, it indicates that the component acts in the direction opposite to that which was assumed.

- See Examples13-1 to 13-5 in text.

13.5 EQUATIONS OF MOTION: NORMAL AND TANGENTIAL COORDINATES

- When a particle moves over a *known curved path*, the equation of motion for the particle may be written in the tangential, normal and binormal directions giving the following three *scalar* equations of motion:

$$\sum F_t = ma_t,$$
$$\sum F_n = ma_n,$$
$$\sum F_b = 0.$$

(13.4)

Inertial coordinate
system

SOLVING PROBLEMS USING THE EQUATIONS OF MOTION (13.4)

When a problem involves the motion of a particle along a known curved path, normal and tangential coordinates should be considered for the analysis since the acceleration components can be readily formulated. Specifically, for t, n and b coordinates we have the following procedure:

- **Free-Body Diagram**

  - Establish the inertial t, n, b coordinate system at the particle and draw the particle's free-body diagram.
  - The particle's normal acceleration $\mathbf{a}_n$ always acts in the positive n direction.
  - If the tangential direction $\mathbf{a}_t$ is known, assume it acts in the positive t direction.
  - Identify the unknowns in the problem.

- **Equations of Motion**

  - Apply the equations of motion (13.4).

- **Kinematics**

  - Formulate the tangential and normal components of acceleration: i.e., $a_t = \frac{dv}{dt}$ or $a_t = v\frac{dv}{ds}$ and $a_n = \frac{v^2}{\rho}$.
  - If the path is defined as $y = f(x)$, the radius of curvature at the point where the particle is located can be obtained from

$$\rho = \frac{\left[1 + \left(\frac{dy}{dx}\right)^2\right]^{\frac{3}{2}}}{\left|\frac{d^2y}{dx^2}\right|}.$$

13.6 EQUATIONS OF MOTION: CYLINDRICAL COORDINATES

- The equation of motion for the particle may be written in the (cylindrical) r, θ, z directions giving the following three *scalar* equations of motion:

$$\sum F_r = ma_r,$$
$$\sum F_\theta = ma_\theta, \qquad\qquad (13.5)$$
$$\sum F_z = ma_z.$$

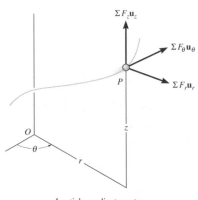

Inertial coordinate system

Note that if the particle is constrained to move only in the $r - \theta$ plane, then only the first two of Eqs. (13.5) are used to specify the motion.

SOLVING PROBLEMS USING THE EQUATIONS OF MOTION (13.5)

Cylindrical or polar coordinates are a suitable choice for the analysis of a problem for which data regarding the angular motion of the radial line r are given, or in cases where the path can be conveniently expressed in terms of these coordinates. Once these coordinates have been established, the equations of motion can be applied in order to relate the forces acting on the particle to its acceleration components. The specific procedure is as follows:

- **Free-Body Diagram**

    - Establish the inertial r, θ, z coordinate system at the particle and draw the particle's free-body diagram.
    - Assume that $\mathbf{a}_r, \mathbf{a}_\theta, \mathbf{a}_z$ act in the positive directions of r, θ, z if they are unknown.
    - Identify the unknowns in the problem.

- **Equations of Motion**

    - Apply the equations of motion (13.5).

- **Kinematics**

    - Determine r and the time derivatives $\dot{r}, \ddot{r}, \dot{\theta}, \ddot{\theta}, \ddot{z}$, and then evaluate the acceleration components $a_r = \ddot{r} - r\dot{\theta}^2$, $a_\theta = r\ddot{\theta} + 2\dot{r}\dot{\theta}$, $a_z = \ddot{z}$.
    - If any of the acceleration components is computed as a negative quantity, it indicates that it acts in its negative coordinate direction.
    - Use the chain rule of calculus to calculate the *time* derivatives of $r = f(\theta)$.

13.7 CENTRAL-FORCE MOTION AND SPACE MECHANICS

If a particle is moving only under the influence of a force having a line of action which is always directed toward a fixed point, the motion is called *central-force motion*. This type of motion is commonly caused by electrostatic and gravitational forces.

- The following differential equation defines the path $r = f(\theta)$ over which the particle travels when it is subjected to the central force **F** (taken positive when it is directed *towards* the fixed point):

$$\frac{d^2\xi}{d\theta^2} + \xi = \frac{F}{mh^2\xi^2} \tag{13.6}$$

where $\xi = \frac{1}{r}$, m is the mass of the particle and h is constant.

- When **F** is the force of *gravitational attraction* e.g., between an artificial satellite and the earth, we have $F = G\frac{M_e m}{r^2}$, where M_e and m represent the mass of the earth and the satellite, respectively, G is the gravitational constant and r is the distance between the mass centres. The solution of (13.6) in this case is

$$\xi = \frac{1}{r} = C\cos(\theta - \phi) + \frac{GM_e}{h^2} \tag{13.7}$$

where C and ϕ are arbitrary (determined from the data obtained for the position and velocity of the satellite). Equation (13.7) represents *free-flight trajectory* of the satellite - notice that Equation (13.7) is the equation of a conic section in polar coordinates.

- The type of (conical) path taken by the satellite is determined from the value of the eccentricity $e = \frac{Ch^2}{GM_e}$ of the conic section described by Equation (13.7):

$$
\begin{aligned}
e &= 0 \quad &&\text{free-flight trajectory is a circle} \\
e &= 1 \quad &&\text{free-flight trajectory is a parabola} \\
e &< 1 \quad &&\text{free-flight trajectory is an ellipse} \\
e &> 1 \quad &&\text{free-flight trajectory is a hyperbola}
\end{aligned}
$$

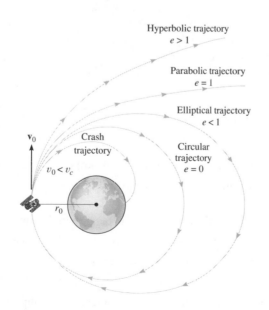

- The initial launch velocity required for the satellite to follow a parabolic path (on the border of never returning to its initial starting point) is called the *escape velocity*. The corresponding speed is given by

$$v_e = \sqrt{\frac{2GM_e}{r_0}}.$$

- The speed v_c required to launch a satellite into a circular orbit is given by

$$v_c = \sqrt{\frac{GM_e}{r_0}}$$

Speeds at launch less than v_c will cause the satellite to reenter the earth's atmosphere and either burn up or crash.

- All trajectories attained by planets and most satellites are *elliptical*. For a satellite's orbit about the earth, the *minimum* distance r_0 from the orbit to the center of the earth is called the *perigee* of the orbit. The *maximum* distance is called the *apogee*.

- In addition to predicting the orbital trajectory of earth satellites, the above theory is also valid in predicting the motion of the planets travelling around the sun (again the solution of Equation (13.6) but with **F** the force of *gravitational attraction* between a planet and the sun, i.e., $F = G\frac{M_s m}{r^2}$, where M_s is the mass of the sun). The fact that planets do indeed follow elliptic orbits about the sun was discovered by Johannes Kepler in the early seventeenth century - after 20 years of planetary observation! This led to *Kepler's laws for planetary motion*.

HELPFUL TIPS AND SUGGESTIONS

- Remember the importance of a free-body diagram in formulating the equation of motion - in any coordinate system.

REVIEW QUESTIONS

1. True or False?

   (a) The equation of motion is not valid without the assumption of an inertial frame.

   (b) The equation of motion is based solely on mathematical arguments.

2. What are the (scalar) equations of motion if a particle is constrained to move only in the $x - y$ plane?

3. When is it preferable for the equation of motion of a particle to be written in normal and tangential coordinates? Why?

4. In Equations (13.4) above, why is there no motion of the particle in the binormal direction?

5. When is it preferable for the equation of motion of a particle to be written in cylindrical coordinates?

6. Do all central force problems result in paths which are conics?

7. In a gravitational central force problem, how do we determine the type of conic describing the path?

8. What method did Johannes Kepler use to discover that planets follow elliptic orbits around the sun?

14

Kinetics of a Particle:
Work and Energy

MAIN GOALS OF THIS CHAPTER:
- To develop the principle of work and energy and apply it to solve problems that involve force, velocity and displacement.
- To study problems that involve power and efficiency.
- To introduce the concept of a conservative force and apply the theorem of conservation of energy to solve kinetic problems.

14.1 THE WORK OF A FORCE

In mechanics a force **F** does work on a particle only when the particle undergoes a *displacement in the direction of the force*.

- The work dU done by the force **F** in displacing a particle **dr** is a scalar quantity defined by

$$dU = \mathbf{F} \cdot \mathbf{dr} = F\,ds\cos\theta$$

  where $ds = |\mathbf{dr}|$ and θ is the angle between the tails of **dr** and **F**.
- **Work of a Variable Force.** If a particle undergoes a finite displacement along its path from $\mathbf{r}_1$ to $\mathbf{r}_2$ or s_1 to s_2 the work done is given by

$$U_{1-2} = \int_{\mathbf{r}_1}^{\mathbf{r}_2} \mathbf{F} \cdot \mathbf{dr} = \int_{s_1}^{s_2} F\cos\theta\,ds \qquad (14.0)$$

- **Work of a Constant Force Moving Along a Straight Line.** Since both F and θ are constant (straight line path)

$$U_{1-2} = F\cos\theta \int_{s_1}^{s_2} ds$$
$$= F\cos\theta\,(s_2 - s_1)$$

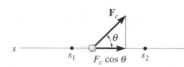

- **Work of a Weight W.**

$$U_{1-2} = -Wy \tag{14.1}$$

where the vertical displacement y is measured *positive upward* (so the work of the weight is *positive* if the particle is displaced *downward* and *negative* if displaced *upward*).

- **Work of a Spring Force.**

  (a) *Work of a Spring.* The work of a spring is of the form $U_s = \frac{1}{2}ks^2$, where k is the spring stiffness and s is the stretch or compression of the spring.

  (b) *Work Done ON the Spring.* If a spring is elongated or compressed from a position s_1 to a further position s_2, the work done on the spring by the force $\mathbf{F}_s$ developed in the (linearly elastic) spring is positive, since, in each case, the force and displacement are in the same direction. Then

$$U_{1-2} = \frac{1}{2}k\left(s_2^2 - s_1^2\right) \tag{14.2}$$

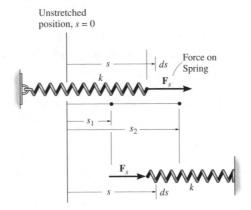

  (c) *Work Done on a Body (or Particle) Attached to a Spring.* In this case, the force $\mathbf{F}_s$ exerted on the body is opposite to that exerted on the spring. Hence, the force $\mathbf{F}_s$ will do *negative* work on the body (particle):

$$U_{1-2} = -\frac{1}{2}k\left(s_2^2 - s_1^2\right) \tag{14.3}$$

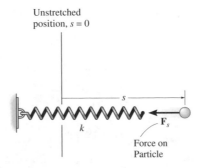

14.2 PRINCIPLE OF WORK AND ENERGY

- The *principle of work and energy for a particle* is described by the equation

$$T_1 + \sum U_{1-2} = T_2 \tag{14.4}$$

where $T_1 = \frac{1}{2}mv_1^2$ and $T_2 = \frac{1}{2}mv_2^2$ represent the kinetic energy of the particle at point1 and point2, respectively, while $\sum U_{1-2}$ represents the work done by all the forces acting on the particle as the particle moves from point 1 to point 2..

- The principle of work and energy is used to solve kinetic problems that involve *velocity, force and displacement* (since these terms are involved in the equation describing the principle i.e., Equation (14.4)).

SOLVING PROBLEMS USING THE PRINCIPLE OF WORK AND ENERGY

- **Work (Free-Body Diagram)**

  - Establish the inertial coordinate system and draw a free-body diagram of the particle in order to account for all the forces that do work on the particle as it moves along its path.

- **Principle of Work and Energy**

  - Apply the principle of work and energy, $T_1 + \sum U_{1-2} = T_2$.
  - The kinetic energy at the initial and final points is always positive, since it involves the speed squared ($T = \frac{1}{2}mv^2$).
  - A force does work when it moves through a displacement in the direction of the force.
  - Work is *positive* when the force component is in the *same direction* as its displacement, otherwise it is negative.
  - Forces that are functions of displacement must be integrated to obtain the work (see Equation (14.0)). Graphically, the work is equal to the area under the force-displacement curve.
  - The work of a weight is the product of the weight magnitude and the vertical displacement, $U_W = \pm Wy$ (see Equation (14.1)). It is positive when the weight moves downwards and negative when the weight moves upwards.
  - The work of a spring is of the form $U_s = \frac{1}{2}ks^2$, where k is the spring stiffness and s is the stretch or compression of the spring (see also Equations (14.2) and (14.3).

14.3 PRINCIPLE OF WORK AND ENERGY FOR A SYSTEM OF PARTICLES

- The *principle of work and energy* can be extended to include a *system of particles* isolated within an enclosed region of space. Symbolically, the principle looks like

$$\sum T_1 + \sum U_{1-2} = \sum T_2. \tag{14.5}$$

In words, this equations states that

the system's initial kinetic energy ($\sum T_1$) plus the work done by all the external and internal forces acting on the particles of the system ($\sum U_{1-2}$) is equal to the system's final kinetic energy ($\sum T_2$).

Note that although the internal forces on adjacent particles occur in equal but opposite collinear pairs, the total work done by each of these forces will, in general, *not cancel* out since the *paths* over which corresponding particles travel will be *different*. There are two important exceptions to this rule which often occur in practice:

- – **When Particles are Contained Within the Boundary of a Translating Rigid Body**.
- – **When Particles are Connected by Inextensible Cables.** In these cases, adjacent particles exert equal but opposite internal forces that have components which undergo the same displacement, and therefore the work of these forces cancels.

- **Special Class of Problems Involving Work of Friction Caused by Sliding**. We note also that Equation (14.5) can be applied to problems involving *sliding friction;* however, it should be realized that the work of the resultant frictional force is not represented by $\mu_k N s$; instead, this term represents both the external work of friction $(\mu_k N s')$ and internal work $(\mu_k N(s - s'))$ which is converted into various forms of internal energy, such as heat.

14.4 POWER AND EFFICIENCY

- **Power.** *Power* is defined as the amount of work performed per unit of time. Hence, the power P generated by a machine or engine which performs an amount of work dU within a time interval dt, is given by

$$P = \frac{dU}{dt}$$
$$= \mathbf{F} \cdot \frac{d\mathbf{r}}{dt}$$
$$= \mathbf{F} \cdot \mathbf{v}$$

where $\mathbf{v}$ is the velocity of the point which is acted upon by the force $\mathbf{F}$.

- – Consequently power is a scalar with basic units *watt (W)* in the SI system and *horsepower (hp)* in the FPS system.

$$1W = 1J/s = 1N \cdot m/s$$
$$1hp = 550 ft \cdot lb/s$$
$$1hp = 746W$$

- **Efficiency.** The *mechanical efficiency of a machine* is defined by

$$\epsilon = \frac{\text{power output}}{\text{power input}}$$

or

$$\epsilon = \frac{\text{energy output}}{\text{energy input}}$$

Since machines consist of a series of moving parts, frictional forces will always be developed within the machine. As a result, extra energy or power is needed to overcome these forces. Consequently, *the efficiency of a machine is always less than one.*

COMPUTING THE POWER SUPPLIED TO A BODY

- First determine the external force $\mathbf{F}$ acting on the body which causes the motion. This force is usually developed by a machine or engine placed either within or external to the body.
- If the body is accelerating, it may be necessary to draw its free-body diagram and apply the equation of motion ($\sum \mathbf{F} = m\mathbf{a}$) to determine $\mathbf{F}$.
- Once $\mathbf{F}$ and the velocity $\mathbf{v}$ of the point where $\mathbf{F}$ is applied have been found, the power is determined by multiplying the force magnitude by the component of velocity acting in the direction of $\mathbf{F}$ i.e., $P = \mathbf{F} \cdot \mathbf{v} = Fv \cos\theta$.
- In some problems, the power may be found by calculating the work done by $\mathbf{F}$ per unit of time ($P_{avg} = \frac{\Delta U}{\Delta t}$ or $P = \frac{dU}{dt}$).

14.5 CONSERVATIVE FORCES AND POTENTIAL ENERGY

- **Conservative Force.** When the work done by a force in moving a particle from one point to another is *independent of the path* followed by the particle, then this force is called a *conservative force.* e.g.,

  - The work done by the *weight of a particle* is *independent of the path* of the particle i.e., the work done depends only on particle's *vertical displacement.*

  - The work done by a spring force *acting on a particle* is *independent of the path* of the particle i.e., it depends only on the extension or compression of the spring.

  - In contrast, we note that the *force of friction* exerted *on a moving object* by a fixed surface *depends on the path* of the object i.e., the longer the path, the greater the work. Consequently, frictional forces are *nonconservative*. The work is dissipated from the body in the form of heat.

- **Potential Energy.** *Potential energy* is a measure of the amount of work a conservative force will do when it moves from a given position to a reference datum.

  - **Gravitational Potential Energy.** The gravitational potential energy of a particle of weight W is

  $$V_g = Wy,$$

  where y is the position of the particle measured *positive upward* from an arbitrarily selected datum.

  - **Elastic Potential Energy.** The elastic potential energy due to a spring's configuration (stretched or compressed a distance s from its unstretched position) is

  $$V_e = \frac{1}{2}ks^2$$

  Note that V_e is *always positive* since, in the deformed position, the force of the spring has the capacity for always doing positive work on the particle when the spring is returned to its unstretched position.

- **Potential Function.** In general, if a particle is subjected to both gravitational and elastic forces, the particle's potential energy can be expressed as a *potential function*, which is the algebraic sum

  $$V = V_g + V_e.$$

- **Proving a Force F is Conservative**. In fact, a force **F** is related to its potential function V by the equation

  $$\mathbf{F} = -\nabla V \tag{14.6}$$

  In other words if a force **F** *and its potential function V*
  satisfy Equation (14.6), then **F** *is a conservative force.*

14.6 CONSERVATION OF ENERGY

- If *only conservative forces* are applied to a body, the principle of work and energy becomes the principle of *conservation of (mechanical) energy* described by:

  $$T_1 + V_1 = T_2 + V_2. \tag{14.7}$$

  In other words, during the motion, the sum of the particle's kinetic and potential energies remains constant (i.e., kinetic energy must be transformed into potential energy and vice versa during the motion).

- **System of Particles**. If a system of particles is subjected to *only conservative forces*, the equation of conservation of energy for the system is

  $$\sum T_1 + \sum V_1 = \sum T_2 + \sum V_2 \tag{14.8}$$

SOLVING PROBLEMS USING THE CONSERVATION OF ENERGY (14.7)

The conservation of energy equation is used to solve problems involving *velocity, displacement and conservative force systems*. It is generally *easier to apply* than the principle of work and energy because the energy equation just requires specifying the particle's kinetic and potential energies at only *two points* along the path, rather than determining the work done when the particle moves through a displacement. The procedure is as follows:

- **Potential Energy**
    - Draw two diagrams showing the particle located at its initial and final points along the path.
    - If the particle is subjected to a vertical displacement, establish the fixed horizontal datum from which to measure the particle's gravitational potential energy V_g.
    - Data pertaining to the elevation y of the particle from the datum and the extension or compression s of any connecting springs can be determined from the geometry associated with the two diagrams.
    - Recall $V_g = Wy$, where y is positive upward from the datum and negative downward from the datum; also $V_e = \frac{1}{2}ks^2$.

- **Conservation of Energy**
    - Apply the equation $T_1 + V_1 = T_2 + V_2$.
    - When determining the kinetic energy, $T = \frac{1}{2}mv^2$, the particle's speed v must be measured from an inertial reference frame.

HELPFUL TIPS AND SUGGESTIONS

- Only problems involving *conservative forces* (weights and springs) may be solved using *conservation of energy* (Equations (14.7) or (14.8)). Friction or other drag-resistant forces, which depend on velocity or acceleration are nonconservative (a portion of the work done by such forces is transformed into thermal energy which dissipates into the surroundings and may not be recovered). When such forces enter into the problem, use the *principle of work and energy*.

REVIEW QUESTIONS

1. How would you calculate the work done by a force $\mathbf{F}$?
2. When is the principle of work and energy used to solve kinetic problems?
3. How is power defined and how is it calculated?
4. Why is the efficiency of a machine always less than one?
5. What is a conservative force? Give some examples of conservative forces.
6. Explain why the weight of an object is a conservative force.
7. Give an example of a nonconservative force and explain why the force is nonconservative.
8. What is potential energy? Give some examples.
9. How can you prove that a force is conservative?
10. When is the conservation of energy equation used to solve problems in kinetics?

15

Kinetics of a Particle:
Impulse and Momentum

MAIN GOALS OF THIS CHAPTER:

- To develop the principle of linear impulse and momentum for a particle.
- To study the conservation of linear momentum for particles.
- To analyze the mechanics of impact.
- To introduce the concept of angular impulse and momentum.
- To solve problems involving steady fluid streams and propulsion with variable mass.

15.1 PRINCIPLE OF LINEAR IMPULSE AND MOMENTUM

- The *principle of linear impulse and momentum* is obtained from a time integration of the equation of motion and is described by the equation

$$m\mathbf{v}_1 + \sum \int_{t_1}^{t_2} \mathbf{F}\, dt = m\mathbf{v}_2. \tag{15.0}$$

  - **Linear Momentum.** Each of the two *vectors* of the form $\mathbf{L} = m\mathbf{v}$ is referred to as the particle's *linear momentum*. It's magnitude is mv and its direction is the same as that of the velocity $\mathbf{v}$.
  - **Linear Impulse.** The integral $\mathbf{I} = \int \mathbf{F}\, dt$ is referred to as the *linear impulse*. This is a *vector quantity* which measures the effect of a force during the time the force acts. The impulse acts in the same direction as the force $\mathbf{F}$ and has units of force-time e.g., $N \cdot s$ or $lb \cdot s$.

- The *principle of linear impulse and momentum* (15.0) states that the initial momentum of the particle plus the sum of all the impulses applied to the particle is equivalent to the final momentum of the particle. It provides a *direct means* of obtaining the particle's final velocity $\mathbf{v}_2$ after a specified time period when the particle's initial velocity is known and the forces acting on the particle are either constant or can be expressed *as functions of time*.

- **Principle of Linear Impulse and Momentum in Scalar Form.**

$$m\,(v_x)_1 + \sum \int_{t_1}^{t_2} F_x\,dt = m\,(v_x)_2\,,$$

$$m\,(v_y)_1 + \sum \int_{t_1}^{t_2} F_y\,dt = m\,(v_y)_2\,, \qquad (15.1)$$

$$m\,(v_z)_1 + \sum \int_{t_1}^{t_2} F_z\,dt = m\,(v_z)_2\,.$$

PROCEDURE FOR SOLVING PROBLEMS

The *principle of linear impulse and momentum* is used to solve problems involving *force, time and velocity.*

- **Free-Body Diagram**

  - Establish the x, y, z inertial frame of reference and draw the particle's free-body diagram in order to account for all the forces that produce impulses on the particle.
  - The direction and sense of the particle's initial and final velocities should be established.
  - If a vector is unknown, assume that the sense of its components is in the direction of the positive inertial coordinates.
  - As an alternative procedure, draw the impulse and momentum diagrams for the particle.

- **Principle of Linear Impulse and Momentum**

  - In accordance with the established coordinate system apply the principle (15.0). If motion occurs in the $x-y$ plane, the two scalar component equations (see Equations (15.1)) can be formulated by either resolving the vector components of F from the free-body diagram or by using the data on the impulse and momentum diagrams.
  - Realize that all forces acting on the particle's free-body diagram will create an impulse, even though some of these forces will do no work.
  - Forces that are functions of time must be integrated to obtain the impulse.
  - If the problem involves the dependent motion of several particles, use Section 12.9 to relate their velocities. Make sure the positive coordinate directions used for writing these kinematic equations are the *same* as those used for writing the equations of impulse and momentum.

15.2 PRINCIPLE OF LINEAR IMPULSE AND MOMENTUM FOR A SYSTEM OF PARTICLES

- The principle of linear impulse and momentum for a *system of particles* moving relative to an inertial reference is given by

$$\sum m_i(\mathbf{v}_i)_1 + \sum \int_{t_1}^{t_2} \mathbf{F}_i\,dt = \sum m_i(\mathbf{v}_i)_2 \qquad (15.2)$$

which states that the initial linear momenta of the system plus the impulses of all the *external forces* acting on the system from t_1 to t_2 are equal to the system's final linear momentum.

- Alternatively, in terms of the *mass center G* of the system, we have

$$m(\mathbf{v}_G)_1 + \sum \int_{t_1}^{t_2} \mathbf{F}\,dt = m(\mathbf{v}_G)_2. \qquad (15.3)$$

15.3 CONSERVATION OF LINEAR MOMENTUM FOR A SYSTEM OF PARTICLES

- When the sum of the external impulses acting on a system of particles is zero, Equation (15.2) becomes

$$\sum m_i(\mathbf{v}_i)_1 = \sum m_i(\mathbf{v}_i)_2 \tag{15.4}$$

which expresses *conservation of linear momentum* i.e., that the linear momenta for a system of particles remain constant during the time period t_1 to t_2.

- In terms of the *mass center G* of the system, we have

$$(\mathbf{v}_G)_1 = (\mathbf{v}_G)_2.$$

That is, the velocity $\mathbf{v}_G$ of the mass center for the system of particles does not change when no external impulses are applied to the system.

- For application, a careful study of the free-body diagram for the *entire* system of particles should be made to identify the forces which create external impulses and thereby determine *in which direction* linear momentum is conserved.

  - If the time period over which the motion is studied is *very short*, some of the external impulses may also be *neglected* or considered approximately equal to zero. The forces causing these negligible impulses are called *nonimpulsive forces* e.g., weight of a body or any force which is very small compared to other larger (impulsive) forces (which change the system's momentum drastically). When making the *distinction* between impulsive and nonimpulsive forces, it is important to realize that this applies only during the (*very short*) time interval t_1 to t_2.

PROCEDURE FOR SOLVING PROBLEMS

Generally, the *principle of linear impulse and momentum* or the *conservation of linear momentum* is applied to a *system of particles* in order to determine the final velocities of the particles *just after* the time period considered. By applying these equations to the entire system, the internal impulses acting within the system, which may be unknown, are *eliminated* from the analysis.

- **Free-Body Diagram**

  - Establish the x, y, z inertial frame of reference and draw the free-body diagram for each particle of the system in order to identify the internal and external forces.
  - The conservation of linear momentum applies to the system *in a given direction* when no external forces or if nonimpulsive forces act on the system *in that direction*.
  - Establish the direction and sense of the particles' initial and final velocities. If the sense is unknown, assume it is along a positive inertial coordinate axis..
  - As an alternative procedure, draw the impulse and momentum diagrams for each particle of the system.

- **Momentum Equations**

  - Apply the principle of linear impulse and momentum or the conservation of linear momentum in the appropriate directions.
  - If it is necessary to determine the *internal impulse* $\int \mathbf{F}\,dt$ acting on only one particle of a system, then the particle must be *isolated* (free-body diagram), and the principle of linear impulse and momentum must be applied *to the particle*.
  - After the impulse is calculated, and provided the time Δt for which the impulse acts is known, the average impulsive force $\mathbf{F}_{avg}$ can be determined from $\mathbf{F}_{avg} = \dfrac{\int \mathbf{F}\,dt}{\Delta t}$.

- See Examples 15.4 to 15.8 in text.

15.4 IMPACT

Impact occurs when two bodies collide with each other during a very short period of time, causing relatively large (impulsive) forces to be exerted between the bodies. In general, there are two types of impact:

- **Central Impact** - when the direction of motion of the mass centers of the two colliding particles is along a line (*line of impact*) passing through the mass centers of the particles.
- **Oblique Impact** - when the motion of one or both of the particles is at an angle with the line of impact.

COEFFICIENT OF RESTITUTION

The coefficient of restitution is defined as the ratio of the relative velocity of the particles' separation *just after impact*, to the relative velocity of the particles' approach *just before impact* i.e.,

$$e = \frac{(v_B)_2 - (v_A)_2}{(v_A)_1 - (v_B)_1}.$$

- **Elastic Impact**: $e = 1$. Deformation impulse is equal and opposite to restitution impulse.
- **Plastic Impact**: $e = 0$. No restitution impulse so that after collision, both particles *stick together* and move with a common velocity.

PROCEDURE FOR SOLVING PROBLEMS (Central Impact)

In most cases, the *final velocities* of two smooth particles are to be determined *just after* they are subjected to direct central impact. Provided the coefficient of restitution, the mass of each particle, and each particle's initial velocity *just before* impact are known, the solution to the problem can be obtained using the following two equations:

- The conservation of momentum applies to the system of particles:

$$\sum m v_1 = \sum m v_2.$$

- The coefficient of restitution e relates the relative velocities of the particles along the ,line of impact, just before and just after collision.

PROCEDURE FOR SOLVING PROBLEMS (Oblique Impact)

When oblique impact occurs between two smooth particles, the particles move away from each other with velocities having *unknown directions* as well as unknown magnitudes. Provided the initial velocities are known, *four* unknowns are present in the problem. These unknowns may be represented either as $(v_A)_2$, $(v_B)_2$, θ_2, and ϕ_2, or as the x and y components of the final velocities.

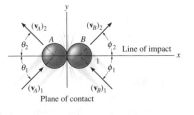
Plane of contact

If the $y - axis$ is established within the plane of contact and the $x - axis$ along the line of impact, the impulsive forces of deformation and restitution *act only in the x direction*. Resolving the velocity or momentum vectors into components along the x and y axes, it is possible to write four independent scalar equations in order to determine $(v_{Ax})_2$, $\left(v_{Ay}\right)_2$, $(v_{Bx})_2$ and $\left(v_{By}\right)_2$.

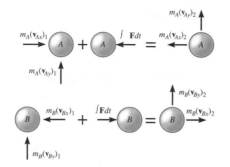

- Momentum of the system is conserved along the line of impact $x - axis$, so that:

$$\sum m(v_x)_1 = \sum m(v_x)_2.$$

- The coefficient of restitution $e = \dfrac{(v_{Bx})_2 - (v_{Ax})_2}{(v_{Ax})_1 - (v_{Bx})_1}$ relates the relative-velocity *components* of the particles *along the line of impact* ($x - axis$).
- Momentum of particle A is conserved along the $y - axis$, perpendicular to the line of impact, since no impulse acts on particle A in this direction.
- Momentum of particle B is conserved along the $y - axis$, perpendicular to the line of impact, since no impulse acts on particle B in this direction.
- **NOTE.** The *principle of work and energy* cannot be used to analyse impact problems since it is not possible to know how the *internal forces* of deformation and restitution vary or displace during the collision.

15.5 ANGULAR MOMENTUM

> The angular momentum $\mathbf{H}_O$ *of a particle about point O is defined as the "moment"*
> *of the particle's linear momentum about O. It is sometimes referred to as the moment of momentum.*

- **Scalar Formulation.** The magnitude of $\mathbf{H}_O$ is given by

$$(H_O)_z = (d)(mv).$$

  - d is the moment arm (perpendicular distance from O to the line of action of $m\mathbf{v}$).
  - *Direction* of $\mathbf{H}_O$ is defined by the right-hand rule.
  - Units for $(H_O)_z$ are $kg \cdot m^2/s$ or $slug \cdot ft^2/s$.

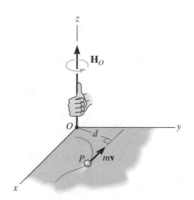

- **Vector Formulation**. If the particle P is moving along a space curve and $\mathbf{r}$ is a position vector drawn from point O to the particle P:

$$\mathbf{H}_O = \mathbf{r} \times m\mathbf{v}.$$

Here the vector $\mathbf{H}_O$ is perpendicular to the plane containing $\mathbf{r}$ and $m\mathbf{v}$.

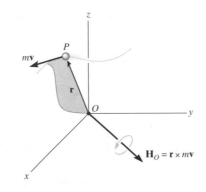

15.6 RELATION BETWEEN MOMENT OF A FORCE AND ANGULAR MOMENTUM

- The resultant moment about point O of all the forces acting on a particle is equal to the time rate of change of the particle's angular momentum about point O i.e.,

$$\sum \mathbf{M}_O = \dot{\mathbf{H}}_O$$

This equation is just *Newton's second law of motion 'for moments'*.

- **System of Particles**. The sum of the moments about point O of all the external forces acting on a system of particles is equal to the time rate of change of the total angular momentum of the system about point O. The point O can be any fixed point in the inertial frame of reference i.e.,

$$\sum (\mathbf{r}_i \times \mathbf{F}_i) = \sum (\dot{\mathbf{H}}_i)_O.$$

15.7 ANGULAR IMPULSE AND MOMENTUM PRINCIPLES

- **Principle of Angular Impulse and Momentum**.

$$(\mathbf{H}_O)_1 + \sum \int_{t_1}^{t_2} \mathbf{M}_O \, dt = (\mathbf{H}_O)_2 \tag{15.5}$$

where the *angular impulse* is defined by

$$\int_{t_1}^{t_2} \mathbf{M}_O \, dt = \int_{t_1}^{t_2} (\mathbf{r} \times \mathbf{F}) \, dt.$$

Here $\mathbf{r}$ is a position vector which extends from any point O to any point on the line of action of the external force $\mathbf{F}$ acting on the particle.

- **Principle of Angular Impulse and Momentum for a System of Particles**

$$\sum (\mathbf{H}_O)_1 + \sum \int_{t_1}^{t_2} \mathbf{M}_O \, dt = \sum (\mathbf{H}_O)_2$$

where again, the angular impulse is created by the moments of the external forces acting on the system.

Impulse and Momentum Principles

- Impulse and momentum principles lead to the following two vector equations describing the particle's motion

$$m\mathbf{v}_1 + \sum \int_{t_1}^{t_2} \mathbf{F} \, dt = m\mathbf{v}_2$$

$$(\mathbf{H}_O)_1 + \sum \int_{t_1}^{t_2} \mathbf{M}_O \, dt = (\mathbf{H}_O)_2 \qquad (15.6)$$

- If the particle is confined to move in the $x - y$ plane, these two vector equations can be written as three scalar equations:

$$m\,(v_x)_1 + \sum \int_{t_1}^{t_2} F_x \, dt = m\,(v_x)_2 \,,$$

$$m\left(v_y\right)_1 + \sum \int_{t_1}^{t_2} F_y \, dt = m\left(v_y\right)_2 \,,$$

$$(H_O)_1 + \sum \int_{t_1}^{t_2} M_O \, dt = (H_O)_2 \,.$$

The first two of these equations represent the principle of linear impulse and momentum in the x and y directions. The third equation represents the principle of angular impulse and momentum about the $z - axis$.

Conservation of Angular Momentum

- When the angular impulses acting on a particle are all zero during the time t_1 to t_2, Equation (15.5) reduces to

$$(\mathbf{H}_O)_1 = (\mathbf{H}_O)_2 \qquad (15.7)$$

which is known as *conservation of angular momentum.*
- If no external impulse is applied to the particle, both linear and angular momentum will be conserved. In some cases, however, the particles angular momentum will be conserved but the linear momentum *will not.* e.g., when the particle is subjected to *only a central force.*
- *Conservation of angular momentum for a system of particles* is given by

$$\sum (\mathbf{H}_O)_1 = \sum (\mathbf{H}_O)_2$$

where the summation must include the angular momenta of *all* particles in the system.

PROCEDURE FOR SOLVING PROBLEMS

When applying the principles of angular impulse and momentum, or the conservation of angular momentum the following procedure should be used:

- **Free-Body Diagram**
  - Draw the particle's free-body diagram in order to identify any axis about which angular momentum may be conserved. For this to occur, the moments of the forces (or impulses) must be parallel or pass through the axis so as to create zero moment throughout the time period t_1 to t_2.
  - The direction and sense of the particles' initial and final velocities should also be established.
  - As an alternative procedure, draw the impulse and momentum diagrams for the particle.

- **Momentum Equations**

  – Apply the principle of angular impulse and momentum (Equation 15.6) or, if appropriate, the conservation of angular momentum (Equation (15.7).

- See Examples 15.13 to 15.15 in text.

15.8 STEADY FLUID STREAMS

Problems involving steady flow can be solved using the following procedure:

- **Kinematic Diagram**

  – If the device is *moving*, a *kinematic diagram* may be helpful for determining the entrance and exit velocities of the fluid flowing onto the device, since a *relative-motion* analysis of velocity will be involved.

  – The measurement of velocities $\mathbf{v}_A$ and $\mathbf{v}_B$ (velocity at which the fluid enters and exits, respectively) must be made by an observer fixed in an inertial frame of reference.

  – Once the velocity of the fluid flowing onto the device is determined, the mass flow $\frac{dm}{dt}$ is calculated using

  $$\frac{dm}{dt} = \rho_A v_A A_A = \rho_B v_B A_B = \rho_A Q_A = \rho_B Q_B.$$

- **Free-Body Diagram**

  – Draw a free-body diagram of the device which is directing the fluid in order to establish the forces $\sum \mathbf{F}$ that act on it. These external forces will include the support reactions, the weight of the device and the fluid contained within it, and the static pressure forces of the fluid at the entrance and exit sections of the device.

- **Equations of Steady Flow**

  – Apply the equations of steady flow:

  $$\sum F_x = \frac{dm}{dt}\left(v_{Bx} - v_{Ax}\right),$$
  $$\sum F_y = \frac{dm}{dt}\left(v_{By} - v_{Ay}\right),$$
  $$\sum M_O = \frac{dm}{dt}\left(d_{OB}v_B - d_{OA}v_A\right),$$

  using the appropriate components of velocity and force shown on the kinematic and free-body diagrams.

- **Examples**: See Examples 15-16 to 15-17 in text.

15.9 PROPULSION WITH VARIABLE MASS

In Section 15.8, we considered the case in which a *constant* amount of mass enters and leaves a *closed* system. In this section, we consider two other cases involving mass flow: these are represented by a system which is either *gaining or losing mass*.

1. **A System that Loses Mass.** Consider a device which at an instant in time has a mass m and is moving forward with a velocity $\mathbf{v}$. At the same time the system is expelling an amount of mass m_e with a mass flow velocity $\mathbf{v}_e$. The governing equation is:

$$\sum F_s = m\frac{dv}{dt} - v_{D/e}\frac{dm_e}{dt}. \tag{15.8}$$

Here, $\sum \mathbf{F}_s$ represents the resultant of all the external forces that *act on the system* in the direction of motion. This *does not include* the force which causes the device to move forward, since this force (*thrust*) is *internal* to the system; $v_{D/e}$ is the magnitude of the relative velocity of the device as seen by an observer moving with the particles of the ejected mass; $\dfrac{dm_e}{dt}$ represents the rate at which mass is being ejected.

2. **A System that Gains Mass.** A device such as a scoop or a shovel may gain mass as it moves forward. Consider a device which at an instant in time has a mass m and is moving forward with a velocity $\mathbf{v}$. At this instant, the device is collecting a particle stream of mass m_i. The flow velocity $\mathbf{v}_i$ of this injected mass is constant and independent of the velocity $\mathbf{v}$ such that $v > v_i$. The governing equation is:

$$\sum F_s = m\frac{dv}{dt} + v_{D/i}\frac{dm_i}{dt}. \qquad (15.9)$$

Here, $\sum \mathbf{F}_s$ represents the resultant of all the external forces that *act on the system* in the direction of motion. This *does not include* the retarding force of the injected mass acting on the device since this force is *internal* to the system; $v_{D/i}$ is the magnitude of the relative velocity of the device as seen by an observer moving with the particles of the injected mass; $\dfrac{dm_i}{dt}$ represents the rate at which mass is being injected into the device.

Note Problems solved using Equations (15.8) and (15.9) should be accompanied by the necessary corresponding free-body diagram. With this diagram, one can then determine $\sum F_s$ *for the system* and isolate the force exerted on the device by the particle stream.

HELPFUL TIPS AND SUGGESTIONS

- Unlike energy, momentum *is a vector* and so has both magnitude and direction. *Impulse and momentum diagrams* will help you keep track of a particle's initial and final momenta for use in the principle of impulse and momentum.

- Remember that when making the distinction between impulsive and nonimpulsive forces, it is important to realize that this applies only in the very short time interval $[t_1, t_2]$ during which the impulse acts.

REVIEW QUESTIONS

1. What is the linear momentum of a particle?

2. When is the principle of *linear impulse and momentum* used to solve problems?

3. When does the principle of linear impulse and momentum become *conservation of linear momentum*?

4. What is a nonimpulsive force?

5. What is the procedure for determining the *final velocities* of two smooth particles *just after* they are subjected to direct central impact if the coefficient of restitution, the mass of each particle, and each particle's initial velocity *just before* impact are known?

6. Define the *angular momentum* of a particle about a point O.

7. What is meant by conservation of angular momentum?

8. True or false? If a particle's angular momentum is conserved the particle's linear momentum must also be conserved.

9. Write down three scalar equations describing impulse and momentum principles for a particle confined to move in the $x - y$ plane.

10. In Equation (15.9) for a system that gains mass, $\sum \mathbf{F}_s$ does not include the retarding force of the injected mass acting on the device. Why?

16

Planar Kinematics of a Rigid Body

MAIN GOALS OF THIS CHAPTER:

- To classify the various types of rigid-body planar motion.

- To investigate rigid-body translation and show how to analyze motion about a fixed axis.

- To study planar motion using an absolute motion analysis.

- To show how to find the instantaneous center of zero velocity and determine the velocity of a point on a body using this method.

- To provide a relative motion analysis of velocity and acceleration using a rotating frame of reference.

16.1 PLANAR RIGID-BODY MOTION

- When all the particles of a rigid body move along paths which are equidistant from a fixed plane, the body is said to undergo *planar motion*. There are three types of rigid body planar motion:

  1. **Translation** Every line segment in the body remains parallel to its original direction during the motion. Specifically, a body can undergo two types of translation:

     i. **Rectilinear Translation**. All points follow parallel straight-line paths.
     ii. **Curvilinear Translation**. All points follow curved paths that are the same shape and are equidistant from one another.

  2. **Rotation about a Fixed Axis**. All of the particles of the body, except those which lie on the axis of rotation, move along circular paths.

  3. **General Plane Motion**. The body undergoes a combination of translation *and* rotation.

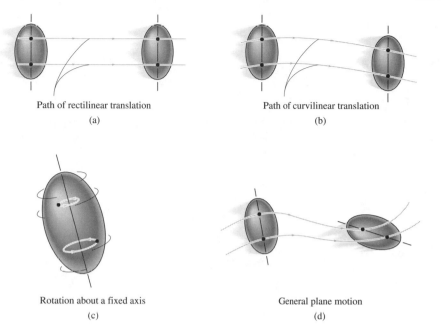

Path of rectilinear translation

(a)

Path of curvilinear translation

(b)

Rotation about a fixed axis

(c)

General plane motion

(d)

16.2 TRANSLATION

Let two points A and B be identified by position vectors $*\mathbf{r_A}$ and $*\mathbf{r_B}$, respectively, from a fixed x, y reference frame. The kinematics of the translating body can be described as follows:

- **Position**

$$\mathbf{r}_B = \mathbf{r}_A + \mathbf{r}_{B/A}$$

where $\mathbf{r}_{B/A}$ is the relative position vector (measures position of B with respect to A).

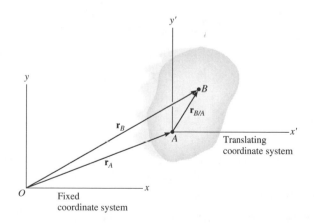

- **Velocity** The body is rigid so the magnitude of $\mathbf{r}_{B/A}$ is constant. The body is translating so the direction of $\mathbf{r}_{B/A}$ is constant. Hence

$$\mathbf{v}_B = \mathbf{v}_A$$

- **Acceleration**

$$\mathbf{a}_B = \mathbf{a}_A$$

In other words, all the points on a translating body move with the same velocity and acceleration.

16.3 ROTATION ABOUT A FIXED AXIS

When a body is rotating about a fixed axis, any point P located in the body travels along a *circular path*. The motion of the *body* is described by *angular motion* which involves three basic quantities: angular position (θ), angular velocity (ω) and angular acceleration (α), described as follows.

- **Angular Motion.**

(a)

- **Angular Velocity**. If θ is the angular position of a radial line r located in some representative plane of the body, the angular velocity ω has magnitude $\omega = \dfrac{d\theta}{dt}$, direction along the axis of rotation and sense of direction either clockwise or counterclockwise. Counterclockwise rotations are usually chosen as *positive*.

- **Angular Acceleration**. The angular acceleration α has magnitude $\alpha = \dfrac{d\omega}{dt} = \dfrac{d^2\theta}{dt^2}$; its sense of direction depends on whether ω is increasing or decreasing (if ω is decreasing, α is called an *angular deceleration* and its sense of direction is opposite to that of ω).

- **Useful Relation Between** α, ω **and** θ. By eliminating t from $\alpha = \dfrac{d\omega}{dt}$ and $\omega = \dfrac{d\theta}{dt}$ we obtain

$$\alpha \, d\theta = \omega \, d\omega$$

– **Constant Angular Acceleration.** If the angular acceleration of the body is constant, $\alpha = \alpha_C$, we have the following set of formulas which relate the body's angular velocity, angular position and time.

$$(\curvearrowright +) \quad \omega = \omega_0 + \alpha_C t$$

$$(\curvearrowright +) \quad \theta = \theta_0 + \omega_0 t + \frac{1}{2}\alpha_C t^2$$

$$(\curvearrowright +) \quad \omega^2 = \omega_0^2 + 2\alpha_C (\theta - \theta_0)$$

Here, θ_0 and ω_0 are the initial values of the body's angular position and angular velocity, respectively, and we have chosen counterclockwise rotations as positive.

- **Motion of Point P**. As the rigid body rotates, point P travels along a circular path of radius r and center at point O. The motion of point P is described by position, velocity and acceleration.

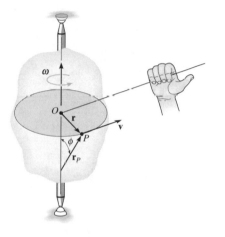

– **Position.** The position of P is defined by the position vector $\mathbf{r}$, which extends from O to P.

– **Velocity.** The velocity $\mathbf{v}$ of P has magnitude $v = r\omega$ and direction always *tangent* to the circular path. This can be represented in vector form as

$$\mathbf{v} = \boldsymbol{\omega} \times \mathbf{r}_P$$

where $\mathbf{r}_P$ is directed from *any* point on the axis of rotation to point P. As a special case, the position vector $\mathbf{r}$ can be chosen for $\mathbf{r}_P$. Here $\mathbf{r}$ lies in the plane of motion and again the velocity of point P is

$$\mathbf{v} = \boldsymbol{\omega} \times \mathbf{r}$$

– **Acceleration**. The acceleration has two components. The *tangential* component of acceleration measures the rate of change in the magnitude of the velocity and can be determined using $a_t = \alpha r$. The *normal* component of acceleration measures the rate of change in the direction of the velocity and can be determined from $a_n = \omega^2 r$. In terms of vectors

$$\mathbf{a}_t = \boldsymbol{\alpha} \times \mathbf{r}$$

$$\mathbf{a}_n = \boldsymbol{\omega} \times (\boldsymbol{\omega} \times \mathbf{r}) = -\omega^2 \mathbf{r}$$

$$\mathbf{a} = \mathbf{a}_t + \mathbf{a}_n = \boldsymbol{\alpha} \times \mathbf{r} - \omega^2 \mathbf{r}$$

PROCEDURE FOR SOLVING PROBLEMS

The velocity and acceleration of a point located on a rigid body that is rotating about a fixed axis can be determined using the following procedure.

- **Angular Motion**

  - Establish the positive sense of direction along the axis of rotation and show it alongside each kinematic equation as it is applied i.e., either ($\curvearrowright$ +) or ($\curvearrowright$ +).

  - If a relationship is known between any two of the four variables α, ω, θ and t, then a third variable can be obtained by using one of the following kinematic equations which relates all three variables

  $$\omega = \frac{d\theta}{dt}, \quad \alpha = \frac{d\omega}{dt}, \quad \alpha d\theta = \omega d\omega.$$

  - If the body's angular acceleration is constant, then the following equations can be used:

  $$\omega = \omega_0 + \alpha_C t$$
  $$\theta = \theta_0 + \omega_0 t + \frac{1}{2}\alpha_C t^2$$
  $$\omega^2 = \omega_0^2 + 2\alpha_C (\theta - \theta_0)$$

  - Once the solution is obtained, the sense of θ, ω and α is determined from the algebraic signs of their numerical quantities.

- **Motion of P**

  - In most cases, the velocity of P and its two components of acceleration can be determined from the scalar equations

  $$v = r\omega, \quad a_t = \alpha r, \quad a_n = \omega^2 r.$$

  - If the geometry of the problem is difficult to visualize, the following vector equations should be used:

  $$\mathbf{v} = \boldsymbol{\omega} \times \mathbf{r}_P = \boldsymbol{\omega} \times \mathbf{r}$$
  $$\mathbf{a}_t = \boldsymbol{\alpha} \times \mathbf{r}_P = \boldsymbol{\alpha} \times \mathbf{r}$$
  $$\mathbf{a}_n = \boldsymbol{\omega} \times (\boldsymbol{\omega} \times \mathbf{r}_P) = -\omega^2 \mathbf{r}$$

  Here, $\mathbf{r}_P$ is directed from any point on the axis of rotation to point P, whereas $\mathbf{r}$ lies in the plane of motion of P. Either of these vectors, along with $\boldsymbol{\omega}$ and $\boldsymbol{\alpha}$ should be expressed in terms of its $\mathbf{i}, \mathbf{j}, \mathbf{k}$ components.

16.4 ABSOLUTE MOTION ANALYSIS

PROCEDURE FOR SOLVING PROBLEMS

The velocity and acceleration of a point P undergoing rectilinear motion can be related to the angular velocity and angular acceleration of a line contained within a body using the following procedure.

- **Position Coordinate Equation**

  - Locate point P using a position coordinate s, which is measured from a *fixed origin* and is directed along the *straight-line path of motion* of point P.

  - Measure from a fixed reference line the angular position θ of a line lying in the body.

  - From the dimensions of the body, relate s to θ, $s = f(\theta)$, using geometry and/or trigonometry.

- **Time Derivatives**

  - Take the first derivative of $s = f(\theta)$ with respect to time to get a relationship between v (speed) and ω.

  - Take the second time derivative to get a relationship between a (magnitude of acceleration) and α.

  - In each case the *chain rule* of calculus must be used when taking the derivatives of the position coordinate equation.

16.5 RELATIVE MOTION ANALYSIS: VELOCITY

PROCEDURE FOR SOLVING PROBLEMS

- The relative velocity equation

$$\mathbf{v}_B = \mathbf{v}_A + \mathbf{v}_{B/A}$$
$$= \mathbf{v}_A + \boldsymbol{\omega} \times \mathbf{r}_{B/A}$$

can be applied either by using Cartesian vector analysis, or by writing the x and y scalar component equations directly.

VECTOR ANALYSIS

- **Kinematic Diagram**

  – Establish the directions of the fixed x and y coordinates and draw a kinematic diagram of the body. Indicate on it the velocities $\mathbf{v}_A$ and $\mathbf{v}_B$ of points A and B, the angular velocity $\boldsymbol{\omega}$ and the relative-position vector $\mathbf{r}_{B/A}$.
  – If the magnitudes of $\mathbf{v}_A$, $\mathbf{v}_B$ or $\boldsymbol{\omega}$ are unknown, the sense of direction of these vectors may be assumed.

- **Velocity Equation**

  – To apply $\mathbf{v}_B = \mathbf{v}_A + \boldsymbol{\omega} \times \mathbf{r}_{B/A}$, express the vectors in Cartesian vector form and substitute them into the equation. Evaluate the cross product and then equate the respective $\mathbf{i}$ and $\mathbf{j}$ components to obtain two scalar equations.
  – If the solution yields a negative answer for an unknown magnitude, it indicates that the sense of direction of the vector is opposite to that shown on the kinetic diagram.

SCALAR ANALYSIS

- **Kinematic Diagram**

  – If the velocity equation is to be applied in scalar form, then the magnitude and direction of the relative velocity $\mathbf{v}_{B/A}$ must be established. Draw a kinetic diagram. Since the body is to be considered pinned momentarily at the base point A, the magnitude is $v_{B/A} = \omega r_{B/A}$. The sense of direction of $\mathbf{v}_{B/A}$ is established from the diagram, such that $\mathbf{v}_{B/A}$ acts perpendicular to $\mathbf{r}_{B/A}$ in accordance with the rotational motion $\boldsymbol{\omega}$ of the body.

- **Velocity Equation**

  – Write the equation $\mathbf{v}_B = \mathbf{v}_A + \mathbf{v}_{B/A}$, and underneath each of the terms, represent the vectors graphically by showing their magnitudes and directions. The scalar equations are determined from the x and y components of these vectors.

16.6 INSTANTANEOUS CENTER OF ZERO VELOCITY

- When using the equation $\mathbf{v}_B = \mathbf{v}_A + \boldsymbol{\omega} \times \mathbf{r}_{B/A}$, the velocity of any point B located on a rigid body can be obtained in a very direct way if one chooses the base point A to be a point that has a *zero velocity* at the instant considered. This point is called the *instantaneous center of zero velocity (IC)*, and it lies on the *instantaneous axis* of zero velocity which is always perpendicular to the plane of motion. Consequently, since, if A is chosen as the IC, $\mathbf{v}_A = \mathbf{v}_{IC} = \mathbf{0}$, and

$$\mathbf{v}_B = \mathbf{v}_{IC} + \boldsymbol{\omega} \times \mathbf{r}_{B/IC}$$
$$= \boldsymbol{\omega} \times \mathbf{r}_{B/IC}$$

Hence, point B moves momentarily about the IC in a *circular path* i.e., the body appears to rotate about the instantaneous axis. For example, for a wheel which rolls without slipping, the point of contact with the ground is an IC.

LOCATION OF THE IC

To locate the *IC* we use the fact that the *velocity* of a point on the body is *always perpendicular* to the relative-position vector extending from the *IC* to the point (since the point is in circular motion about the *IC*). There are three possibilities:

- **Given the velocity v_A of a point *A* on the body and the angular velocity ω of the body.** In this case, the *IC* is located along the line drawn perpendicular to v_A at *A*, such that the distance from *A* to the *IC* is $r_{A/IC} = v_A/\omega$. Note that the *IC* lies up and to the right of *A* since v_A must cause a clockwise angular velocity ω about the *IC*.

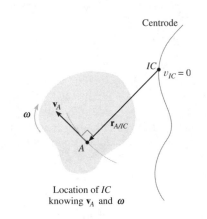

Location of *IC*
knowing $\mathbf{v}_A$ and $\boldsymbol{\omega}$

- **Given the lines of action of two nonparallel velocities v_A and v_B .** Construct at points *A* and *B* line segments that are perpendicular to v_A and v_B. Extending these perpendiculars to their *point of intersection* as shown, locates the *IC* at the instant considered.

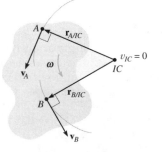

Location of *IC*
knowing the lines of action of $\mathbf{v}_A$ and $\mathbf{v}_B$

- **Given the magnitude and direction of two parallel velocities v_A and v_B.** Here, the location of the IC is determined by proportional triangles.

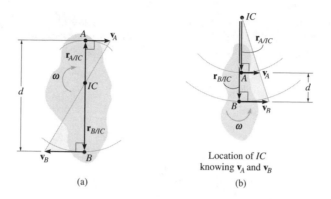

(a)   Location of IC
knowing v_A and v_B
(b)

IMPORTANT NOTES

1. The point chosen as the IC for the body can be used *only for an instant of time* since the body changes its position from one instant to the next.

2. The IC *does not, in general, have zero acceleration* and so *should not* be used for finding accelerations of points in a body.

SOLVING PROBLEMS USING THE IC

- First establish the location of the IC using one of the methods described above.
- The body is then imagined as "extended and pinned" at the IC such that, at the instant considered, it rotates about this pin with its angular velocity ω.
- The magnitude of velocity for an arbitrary point on the body can be determined using the equation $v = r\omega$ where r is the radial line drawn from the IC to each point.
- The line of action of each velocity vector $\mathbf{v}$ is perpendicular to its associated radial line $\mathbf{r}$, and the velocity has a sense of direction which tends to move the point in a manner consistent with the angular rotation ω of the radial line.

16.7 RELATIVE-MOTION ANALYSIS: ACCELERATION

- An equation relating the accelerations of two points on a rigid body subjected to general plane motion is given by

$$\mathbf{a}_B = \mathbf{a}_A + \boldsymbol{\alpha} \times \mathbf{r}_{B/A} - \omega^2 \mathbf{r}_{B/A} \tag{16.0}$$

where

$$
\begin{aligned}
\mathbf{a}_B &= \text{acceleration of base point } B \\
\mathbf{a}_A &= \text{acceleration of base point } A \\
\boldsymbol{\alpha} &= \text{angular acceleration of the body} \\
\omega &= \text{angular velocity of the body} \\
\mathbf{r}_{B/A} &= \text{relative-position vector drawn from } A \text{ to } B
\end{aligned}
$$

• Important Points

1. Before applying Equation (16.0), it will be necessary to determine the angular velocity ω of the body by using a velocity analysis.

2. If Equation (16.0) is applied to a rigid body which is pin-connected to two other bodies, it should be noted that points which are *coincident at the pin* move with the *same acceleration*, since the path of motion over which they travel is the *same*. Fort example, point B lying on either rod AB or BC in the following crank mechanism.

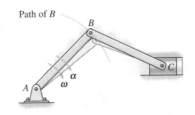

3. If two bodies contact one another *without slipping*, and the *points in contact* move along *different paths*, the *tangential components* of acceleration of the points will be the *same*; however, the *normal components* will *not* be the same. For example, consider point A on gear B and the coincident point A' on gear C. Both move with the same tangential acceleration (due to rotational motion) but since they both follow different curved paths, their normal accelerations are different. Consequently $(a_A)_n \neq (a_{A'})_n$ so that $\mathbf{a}_A \neq \mathbf{a}_{A'}$.

16.8 RELATIVE-MOTION ANALYSIS USING ROTATING AXES

• Consider an x, y, z coordinate system (with origin at point A) which is assumed to be translating and rotating with respect to a fixed X, Y, Z coordinate system. We have the following two equations which describe the velocity and acceleration of a point B.

$$\mathbf{v}_B = \mathbf{v}_A + \mathbf{\Omega} \times \mathbf{r}_{B/A} + \left(\mathbf{v}_{B/A}\right)_{xyz} \tag{16.1}$$

$$\mathbf{a}_B = \mathbf{a}_A + \dot{\mathbf{\Omega}} \times \mathbf{r}_{B/A} + \mathbf{\Omega} \times \left(\mathbf{\Omega} \times \mathbf{r}_{B/A}\right) + 2\mathbf{\Omega} \times \left(\mathbf{v}_{B/A}\right)_{xyz} + \left(\mathbf{a}_{B/A}\right)_{xyz} \tag{16.2}$$

where

$\mathbf{v}_B, \mathbf{a}_B$ = velocity/acceleration of B measured from the X, Y, Z reference

$\mathbf{v}_A, \mathbf{a}_A$ = velocity/acceleration of the origin A of the x, y, z reference, measured from the X, Y, Z reference

$\left(\mathbf{v}_{B/A}\right)_{xyz}, \left(\mathbf{a}_{B/A}\right)_{xyz}$ = relative velocity/acceleration of "B with respect to A," as measured by an observer attached to the x, y, z reference

$\mathbf{\Omega}, \dot{\mathbf{\Omega}}$ = angular velocity/acceleration of the x, y, z reference, measured from the X, Y, Z reference.

$\mathbf{r}_{B/A}$ = relative position of "B with respect to A."

- The term $2\mathbf{\Omega} \times \left(\mathbf{v}_{B/A}\right)_{xyz}$ is called the *Coriolis acceleration*.

Solving Problems Using Equations (16.1) and (16.2)

- **Coordinate Axes**
  - Choose an appropriate location for the origin and proper orientation of the axes for both the X, Y, Z and moving x, y, z reference frames.
  - Most often solutions are easily obtained if at the instant considered:
    1. the origins are coincident
    2. the corresponding axes are collinear
    3. the corresponding axes are parallel
  - The moving frame should be selected fixed to the body or device along which the relative motion occurs.

- **Kinematic Equations**
  - After defining the origin A of the moving reference and specifying the moving point B write Equations (16.1) to (16.2).
  - The Cartesian components of all these vectors may be expressed along either the X, Y, Z axes or the x, y, z axes. The choice is arbitrary provided a consistent set of unit vectors is used.
  - Motion of the moving reference, is expressed by $\mathbf{v}_A$, $\mathbf{a}_A$, $\mathbf{\Omega}$ and $\dot{\mathbf{\Omega}}$; and the motion of B with respect to the moving reference, is expressed by $\mathbf{r}_{B/A}$, $\left(\mathbf{v}_{B/A}\right)_{xyz}$ and $\left(\mathbf{a}_{B/A}\right)_{xyz}$.

HELPFUL TIPS AND SUGGESTIONS

- When using the relative velocity and acceleration equations

$$\mathbf{v}_B = \mathbf{v}_A + \boldsymbol{\omega} \times \mathbf{r}_{B/A}$$
$$\mathbf{a}_B = \mathbf{a}_A + \boldsymbol{\alpha} \times \mathbf{r}_{B/A} - \omega^2 \mathbf{r}_{B/A}$$

the choice of point A is essential. It should always be a point whose velocity/acceleration is *known* or *easy to find*.

- Recall that the IC of zero velocity *does not, in general, have zero acceleration* so that although $\mathbf{v}_{IC} = \mathbf{0}$, $\mathbf{a}_{IC} \neq \mathbf{0}$.

REVIEW QUESTIONS

1. What are the three types of rigid body planar motion? Give a short description of each.
2. Can a point have an angular velocity?
3. In absolute motion analysis, we establish a position-coordinate equation $s = f(\theta)$. How do we then obtain the speed and magnitude of acceleration?
4. If a position-coordinate equation is given by $s = f(\theta) = \sin^2\theta$ where $\theta(t)$ and t is time, find the speed $\frac{ds}{dt}$.
5. For a rigid body in plane motion, is the IC of zero velocity always located either on the body or at a finite distance from the body?
6. Once the IC of zero velocity has been found we can write

$$\mathbf{v}_B = \mathbf{v}_{IC} + \boldsymbol{\omega} \times \mathbf{r}_{B/IC}$$
$$= \boldsymbol{\omega} \times \mathbf{r}_{B/IC}.$$

Is it the case that we can also write

$$\mathbf{a}_B = \boldsymbol{\alpha} \times \mathbf{r}_{B/IC} - \omega^2 \mathbf{r}_{B/IC}?$$

7. What does it mean when the term $\left(\mathbf{v}_{B/A}\right)_{xyz}$ is zero?

17

Planar Kinetics of a Rigid Body: Force and Acceleration

MAIN GOALS OF THIS CHAPTER:

- To introduce the methods used to determine the mass moment of inertia of a body.
- To develop the planar kinetic equations of motion for a symmetric rigid body.
- To discuss applications of these equations to bodies undergoing translation, rotation about a fixed axis, and general plane motion.

17.1 MOMENT OF INERTIA

- The *moment of inertia* is a measure of the resistance of a body to *angular acceleration* in the same way that mass is a measure of the body's resistance to *acceleration*.
- The body's moment of inertia about the $z - axis$ is

$$I = \int_m r^2 \, dm$$

where the "moment arm" r is the perpendicular distance from the $z - axis$ to the arbitrary element dm. Clearly I is always positive and has units of $kg \cdot m^2$ or $slug \cdot ft^2$.

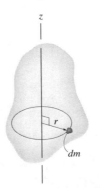

- If the body consists of material having a variable density, $\rho = \rho\,(x, y, z)$, we can write $dm = \rho\,dV$ and express I in terms of volume:

$$I = \int_V r^2 \rho\, dV$$

  If ρ is constant (*homogeneous solid*), we can write

$$I = \rho \int_V r^2\, dV$$

- When the elemental volume chosen for integration has infinitesimal dimensions in all three directions, e.g., $dV = dx\,dy\,dz$, the moment of inertia of the body must be determined using "triple integration." The integration process can, however, be simplified to a *single integration* provided the chosen elemental volume has a differential size or thickness in only *one direction*. Shell or disk elements are often used for this purpose.

- **PARALLEL-AXIS THEOREM.** If the moment of inertia I_G of a body about an axis passing through the body's mass center G is known, then the moment of inertia I about any other parallel axis is given by

$$I = I_G + md^2$$

  where

$$m = \text{mass of body}$$
$$d = \text{perpendicular distance between the parallel axes}$$

- **RADIUS OF GYRATION.** The radius of gyration k (units of length) is defined by

$$I = mk^2 \text{ or } k = \sqrt{\frac{I}{m}}$$

- **COMPOSITE BODIES.** If a body is constructed of a number of simple shapes such as disks, spheres and rods, the moment of inertia of the body about any axis z can be found by adding algebraically the moments of inertia of all the composite shapes computed about the $z - axis$. Here, I_G for each of the composite parts is usually determined from tables (see inside back cover of text).

17.2 PLANAR KINETIC EQUATIONS OF MOTION

- **Equation of Translational Motion.** Measured from an inertial reference frame x, y, z the equation of translational motion is given by

$$\sum \mathbf{F} = m\mathbf{a}_G$$

  or, in scalar form:

$$\sum F_x = m\,(a_G)_x$$
$$\sum F_y = m\,(a_G)_y$$

  where $\sum \mathbf{F}$ is the sum of all external forces acting on the body and $\mathbf{a}_G$ is the acceleration of the body's mass center.

- **Equation of Rotational Motion.** The equation of rotational motion takes two forms:

  i. *When the moments are computed about the body's mass center G :*

$$\curvearrowright \sum M_G = I_G \alpha$$

  where $\sum M_G$ is the sum of the moments of all the external forces and couple moments computed about the point G, I_G is the moment of inertia of the body about an axis passing through G and α is the angular acceleration of the body.

ii. *When the moments are computed about a point* $P \neq G$:

$$\curvearrowright \sum M_P = \sum (\mathcal{M}_k)_P$$

or

$$\curvearrowright \sum M_P = I_G \alpha + \text{ kinetic moments of the components of } m\mathbf{a}_G \text{ about } P$$

where $\sum M_P$ is the sum of the moments of all the external forces and couple moments computed about the point P.

- **Free-Body Diagram.** A good free-body diagram will help you to identify the terms involved in each of $\sum \mathbf{F}$, $\sum M_G$ or $\sum M_P$. A kinetic diagram is especially convenient for identifying the moment terms in $\sum (\mathcal{M}_k)_P$.

17.3 EQUATIONS OF MOTION: TRANSLATION

- **Rectilinear Translation**

$$\sum F_x = m \, (a_G)_x$$

$$\sum F_y = m \, (a_G)_y$$

$$\sum M_G = 0$$

or summing moments about $A \neq G$:

$$\sum M_A = \sum (\mathcal{M}_k)_A$$

Here, $\sum (\mathcal{M}_k)_A$ represents only the moments of the two components of $m\mathbf{a}_G$ about A (since, in translation $I_G \alpha = \mathbf{0}$).

- **Curvilinear Translation**

$$\sum F_n = m \, (a_G)_n$$

$$\sum F_t = m \, (a_G)_t$$

$$\sum M_G = 0$$

or summing moments about $B \neq G$:

$$\sum M_B = \sum (\mathcal{M}_k)_B$$

Here, $\sum (\mathcal{M}_k)_B$ represents only the moments of the components $m \, (\mathbf{a}_G)_n$ and $m \, (\mathbf{a}_G)_t$ about point B (since, in translation $I_G \alpha = \mathbf{0}$).

PROCEDURE FOR SOLVING PROBLEMS

Kinetic problems involving rigid-body translation can be solved using the following procedure:

- **Free-Body Diagram**

  - Establish the x, y or n, t inertial frame of reference and draw the free-body diagram in order to account for all the external forces and couple moments that act on the body.
  - The direction and sense of the acceleration of the body's mass center $\mathbf{a}_G$ should be established.
  - Identify the unknowns in the problem.
  - If it is decided that the rotational equation of motion $\sum M_P = \sum (\mathcal{M}_k)_P$ is to be used, consider drawing the kinetic diagram since it graphically accounts for the components $m \, (\mathbf{a}_G)_x$, $m \, (\mathbf{a}_G)_y$ or $m \, (\mathbf{a}_G)_n$, $m \, (\mathbf{a}_G)_t$ and is therefore convenient for "visualizing" the terms needed in the moment sum $\sum (\mathcal{M}_k)_P$.

- **Equations of Motion**

    - Apply the three equations of motion in accordance with the established sign convention.
    - To simplify the analysis, the moment equation $\sum M_G = 0$ can be replaced by the more general equation $\sum M_P = \sum (\mathcal{M}_k)_P$, where point P is usually located at the intersection of the lines of action of as many unknown forces as possible.
    - If the body is in contact with a rough surface and slipping occurs, use the frictional equation $F = \mu_k N$. Remember, **F** always acts on the body so as to oppose the motion of the body relative to the surface it contacts.

- **Kinematics**

    - Use kinematics if the velocity and position of the body are to be determined.
    - For *rectilinear translation* with *variable acceleration* use

    $$a_G = \frac{dv_G}{dt}, \quad a_G ds_G = v_G dv_G, \quad v_G = \frac{ds_G}{dt}$$

    - For *rectilinear translation* with *constant acceleration*, use

    $$v_G = (v_G)_0 + a_G t, \quad v_G^2 = (v_G)_0^2 + 2a_G[s_G - (s_G)_0],$$
    $$s_G = (s_G)_0 + (v_G)_0 t + \frac{1}{2}a_G t^2$$

    - For *curvilinear translation*, use

    $$(a_G)_n = \frac{v_G^2}{\rho} = \omega^2 \rho, \quad (a_G)_t = \frac{dv_G}{dt}, \quad (a_G)_t ds_G = v_G dv_G,$$
    $$(a_G)_t = \alpha \rho$$

17.4 EQUATIONS OF MOTION: ROTATION ABOUT A FIXED AXIS

- Consider a rigid body which is constrained to rotate in the vertical plane about a fixed axis perpendicular to the page and passing through the pin at O.

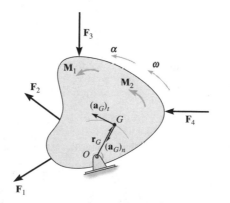

The three equations of motion are:

$$\sum F_n = m\,(a_G)_n = m\omega^2 r_G$$

$$\sum F_t = m\,(a_G)_t = m\alpha r_G$$

$$\sum M_G = I_G\alpha$$

or for a point $O \neq G$

$$\sum M_O = \sum (\mathcal{M}_k)_O = I_O\alpha$$

Note that "$I_O\alpha$" accounts for the "moment" of both $m\,(a_G)_t$ and $I_G\alpha$ about the point O = (the moment of $m\,(a_G)_n$ is not included since the line of action of this vector passes through O).

PROCEDURE FOR SOLVING PROBLEMS

Kinetic problems involving rotation of a body about a fixed axis can be solved using the following procedure:

- **Free-Body Diagram**

    - Establish the x, y or n, t inertial coordinate system and specify the directions and sense of the accelerations $(a_G)_n$ and $(a_G)_t$ and the angular acceleration α of the body. Recall that $(a_G)_t$ must act in a direction which is consistent with α whereas $(a_G)_n$ always acts towards the axis of rotation which is point O.

    - Draw the free-body diagram in order to account for all the external forces and couple moments that act on the body.

    - Compute the moment of inertia I_G or I_O.

    - Identify the unknowns in the problem.

    - If it is decided that the rotational equation of motion $\sum M_P = \sum (\mathcal{M}_k)_P$ is to be used, i.e., P is a point other than G or O, consider drawing the kinetic diagram in order to help "visualize" the "moments" developed by the components $m\,(a_G)_n$, $m\,(a_G)_t$ and $I_G\alpha$ when writing the terms for the moment sum $\sum (\mathcal{M}_k)_P$.

- **Equations of Motion**

    - Apply the three equations of motion in accordance with the established sign convention.

    - If moments are summed about the body's mass center G, then $\sum M_G = I_G\alpha$ since $m\,(a_G)_n$ and $m\,(a_G)_t$ create no moment about G.

    - If moments are summed about the pin support O on the axis of rotation, then $m\,(a_G)_n$ creates no moment about O and it can be shown that $\sum M_O = I_O\alpha$.

- **Kinematics**

    - Use kinematics if a complete solution cannot be obtained strictly from the equations of motion.

    - If *angular acceleration is variable,* use

    $$\alpha = \frac{d\omega}{dt}, \quad \alpha\,d\theta = \omega\,d\theta, \quad \omega = \frac{d\theta}{dt}$$

    - If *angular acceleration is constant,* use

    $$\omega = \omega_0 + \alpha_C t, \quad \omega^2 = \omega_0^2 + 2\alpha_C[\theta - \theta_0],$$

    $$\theta = \theta_0 + \omega_0 t + \frac{1}{2}\alpha_C t^2$$

17.5 EQUATIONS OF MOTION: GENERAL PLANE MOTION

- If an x, y inertial coordinate system is used, the three equations of motion are

$$\sum F_x = m\,(a_G)_x$$
$$\sum F_y = m\,(a_G)_y$$
$$\sum M_G = I_G\alpha$$

or summing moments about $P \neq G$:

$$\sum M_P = \sum (\mathcal{M}_k)_P$$

Here, $\sum (\mathcal{M}_k)_P$ represents the moment sum of $I_G\alpha$ and $m\mathbf{a}_G$ (or its components) about P as determined by the data on the kinetic diagram.

- **Frictional Rolling Problems.** In addition to the three equations of motion for general plane motion, *frictional rolling problems* (involving e.g., wheels, disks, cylinders, or balls) often require an extra equation due to the presence of the 'extra unknown' representing the frictional force. There are two cases:

  - **No slipping.** In this case we have the 'extra equation'

$$a_G = r\alpha.$$

  Note that when the solution is obtained, the assumption of no slipping must be checked (i.e., verify that $F \leq \mu_s N$) otherwise it is necessary to rework the problem under the assumption of *slipping*.

  - **Slipping.** Here, α and a_G are *independent of each other* so instead we relate the magnitude of the frictional force F to the magnitude of the normal force N using the coefficient of kinetic friction μ_k and obtain the 'extra equation'

$$F = \mu_k N$$

PROCEDURE FOR SOLVING PROBLEMS

Kinetic problems involving general plane motion of a rigid-body can be solved using the following procedure:

- **Free-Body Diagram**

  - Establish the x, y inertial frame of reference and draw the free-body diagram in order to account for all the external forces and couple moments that act on the body.
  - The direction and sense of the acceleration of the body's mass center $\mathbf{a}_G$ and the angular acceleration $\boldsymbol{\alpha}$ of the body should be established.
  - Compute the moment of inertia I_G
  - Identify the unknowns in the problem.
  - If it is decided that the rotational equation of motion $\sum M_P = \sum (\mathcal{M}_k)_P$ is to be used, consider drawing the kinetic diagram in order to help "visualize" the "moments" developed by the components $m\,(\mathbf{a}_G)_x$, $m\,(\mathbf{a}_G)_y$ and $I_G\alpha$ when writing the terms for the moment sum $\sum (\mathcal{M}_k)_P$.

- **Equations of Motion**

  - Apply the three equations of motion in accordance with the established sign convention.
  - When friction is present, there is the possibility for motion with no slipping or tipping. Each possibility for motion should be considered.

- **Kinematics**

  - Use kinematics if a complete solution cannot be obtained strictly from the equations of motion.
  - If the body's motion is *constrained* due to its supports, additional equations may be obtained by using $\mathbf{a}_B = \mathbf{a}_A + \mathbf{a}_{B/A}$, which related the accelerations of any two points A and B on the body.
  - When a wheel, disk, cylinder, or ball rolls *without slipping* then $a_G = r\alpha$.

HELPFUL TIPS AND SUGGESTIONS

- Remember always to draw a *free-body diagram* to account for all the external forces and couple moments that act on the body. Also, a good *kinetic diagram* will help you to visualize the "moments" when writing the terms for the moment sum $\sum (\mathcal{M}_k)_P$.
- Before attempting any of the problems, study Examples 17-5 to 17-17 in the text. Re-work a few of these *yourself* knowing you have the full solution available. This is an excellent way to reinforce ideas and understand the relevant material.

REVIEW QUESTIONS

1. What is meant by a moment of inertia?
2. What is meant by a homogeneous solid?
3. What's the easiest way to calculate moments of inertia for common homogeneous solids?
4. In the equation of rotational motion, what is meant by the term $\sum (\mathcal{M}_k)_P$?
5. True or False? In translational problems, the moments of external forces and couples taken about any point add to zero.
6. Is it true that in problems involving rotation about a fixed axis, we can write $\sum M_P = I_P \alpha$ for any point P?
7. Is it true that $\sum M_{IC} = I_{IC}\alpha$ where IC represents the instantaneous center of zero velocity?
8. Consider the homogeneous disk with mass m subjected to a known horizontal force **P**. Draw a free-body diagram and write down three equations of motion for the disk. Next, assuming the disk rolls without slipping, write down another equation for the disk. Is there enough information to solve the problem of finding e.g., the acceleration of G?

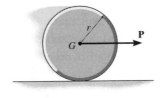

18

Planar Kinetics of a Rigid Body: Work and Energy

MAIN GOALS OF THIS CHAPTER:

- To develop formulations for the kinetic energy of a body and define the various ways a force and couple do work.
- To apply the principle of work and energy to solve rigid-body planar kinetic problems that involve force, velocity and displacement.
- To show how the conservation of energy can be used to solve rigid-body planar kinetic problems.

18.1 KINETIC ENERGY

- **Translation.** When a rigid body of mass m is subjected to either *rectilinear or curvilinear translation*, the kinetic energy of the body is

$$T = \frac{1}{2}mv_G^2$$

where v_G is the magnitude of the translational velocity $\mathbf{v}_G$ at the instant considered. Clearly, in this case, T is made up solely from translational kinetic energy.

- **Rotation about a Fixed Axis.** When a rigid body is *rotating about a fixed axis* passing through point O, the kinetic energy of the body is given by

$$T = \frac{1}{2}mv_G^2 + \frac{1}{2}I_G\omega^2$$
$$= \frac{1}{2}I_O\omega^2$$

Here ω is the angular velocity of the body and I_O is the body's moment of inertia computed about the $z-axis$ passing through point O. In this case, it is clear that T is made up from *both translational kinetic energy* $\frac{1}{2}mv_G^2$ *and rotational kinetic energy about the body's mass center,* $\frac{1}{2}I_G\omega^2$. However, since the motion is one of rotation about O, v_G takes a *special form* ($v_G = r_G\omega$) which, using the parallel axis theorem, leads to the more compact expression $T = \frac{1}{2}I_O\omega^2$.

- **General Plane Motion.** When a rigid body is subjected to general plane motion, its kinetic energy is defined by

$$T = \frac{1}{2}mv_G^2 + \frac{1}{2}I_G\omega^2$$

In this case, T is the *scalar* sum of the body's translational kinetic energy $\frac{1}{2}mv_G^2$ and rotational kinetic energy about its mass center, $\frac{1}{2}I_G\omega^2$. Note that there is now *no special form* for v_G (unlike rotational case) so we *cannot collapse* the expression for T using the parallel axis theorem.

- **System of Connected Rigid Bodies.** Since energy is a scalar quantity, the total kinetic energy for a system of connected rigid bodies is the sum of the kinetic energies of all its moving parts.

18.2 THE WORK OF A FORCE

- **Work of a Variable Force.** If an external force **F** acts on a rigid body, the work done by the force when it moves along the path s is defined as

$$U_F = \int_S F \cos\theta \, ds$$

where θ is the angle between the tails of the force vector and the differential displacement.

- **Work of a Constant Force.** If an external force $\mathbf{F}_c$ acts on a rigid body and maintains a constant magnitude F_c and constant direction θ, the work becomes

$$U_{F_c} = (F_c \cos\theta)\, s$$

Here, $F_c \cos\theta$ represents the magnitude of the component of force in the direction of displacement.

- **Work of a Weight.** The weight of a body does work only when the body's mass center G undergoes a *vertical displacement* $\triangle y$. If this displacement is *upward*, the work is *negative*

$$U_W = -W\triangle y$$

Likewise, if the displacement is *downward* $(-\triangle y)$, the work is *positive*

$$U_W = W\triangle y$$

- **Work of a Spring Force.** If a linearly elastic spring is attached to a body, the spring force $F = ks$ *acting on the body* does work when the spring either stretches or compresses from s_1 to a *further* position s_2. In both cases, the work is negative (since body's displacement is in opposite direction of force):

$$U_s = -\frac{1}{2}k\left(s_2^2 - s_1^2\right), \quad |s_2| > |s_1|$$

- **Forces that do no Work.** There are some external forces that do no work when the body is displaced. These forces can act either at *fixed points* on the body or they can have a *direction perpendicular* to their displacement. e.g., reactions at a pin support about which a body rotates or the weight of a body when the center of gravity of the body moves in a *horizontal plane*.

18.3 THE WORK OF A COUPLE

- When a body subjected to a couple undergoes general plane motion, the two couple forces do work only when the body undergoes a *rotation*. When the body rotates in the plane through a finite angle θ (radians) from θ_1 to θ_2, the work of the couple is then

$$U_M = \int_{\theta_1}^{\theta_2} M \, d\theta$$

- If the couple moment **M** has a *constant magnitude*, then

$$U_M = M(\theta_2 - \theta_1)$$

Here, the work is positive provided **M** and $(\boldsymbol{\theta_2} - \boldsymbol{\theta_1})$ are in the same direction.

18.4 PRINCIPLE OF WORK AND ENERGY

The principle of work and energy for a rigid body is

$$T_1 + \sum U_{1-2} = T_2$$

PROCEDURE FOR SOLVING PROBLEMS

The principle of work and energy is used to solve kinetic problems that involve *velocity, force and displacement,* since these terms are involved in the formulation. We use the following procedure:

- **Work (Free-Body Diagram)**

  - Draw a free-body diagram of the body when it is located at an intermediate point along the path in order to account for all the forces and couple moments which do work on the body as it moves along the path.
  - A force does work when it moves through a displacement *in the direction* of the force.
  - Forces that are functions of displacement must be integrated to obtain the work.
  - Since *algebraic addition* of the work terms is required, it is important that the proper sign of each term be specified. Specifically, work is positive when the force (couple moment) is in the *same direction* as its displacement (rotation); otherwise it is negative.

- **Principle of Work and Energy**

  - Apply the principle of work and energy, $T_1 + \sum U_{1-2} = T_2$. Since this is a scalar equation, it can be used to solve for only one unknown when it is applied to a single rigid body.

18.5 CONSERVATION OF ENERGY

- When a force system acting on a rigid body consists only of *conservative forces*, the conservation of energy theorem $(T_1 + V_1 = T_2 + V_2)$ may be used to solve a problem which would otherwise be solved using the principle of work and energy. This theorem is often easier to apply since the work of a conservative force is independent of the path and depends on only the initial and final positions of the body.

PROCEDURE FOR SOLVING PROBLEMS

The conservation of energy equation is used to solve kinetic problems that involve *velocity, displacement and conservative force systems*. We use the following procedure:

- **Potential Energy**

  - Draw two diagrams showing the body located at its initial and final positions along the path.
  - If the center of gravity G is subjected to a *vertical displacement*, establish a fixed horizontal datum from which to measure the body's gravitational potential energy V_g.
  - Data pertaining to the elevation y_G of the body's center of gravity from the datum and the extension or compression of any connecting springs can be determined from the problem geometry and listed on the two diagrams.
  - Recall that the potential energy $V = V_g + V_e$. Here, $V_g = W y_G$, which can be positive or negative and $V_e = \frac{1}{2}ks^2$, which is always positive..

- **Kinetic Energy**

  - The kinetic energy of the body consists of two parts, namely translational kinetic energy, $T = \frac{1}{2}mv_g^2$ and rotational kinetic energy, $T = \frac{1}{2}I_G\omega^2$.
  - Kinetic diagrams for velocity may be useful for determining v_G and ω or for establishing a *relationship* between these quantities.

- **Conservation of Energy**
  - Apply the conservation of energy equation $T_1 + V_1 = T_2 + V_2$.

HELPFUL TIPS AND SUGGESTIONS

- Remember that unlike force, acceleration or displacement, energy is a *scalar*.
- A brief review of Sections 16.5 to 16.7 may prove helpful in solving problems involving energy since computations for kinetic energy require a kinematic analysis of velocity.
- Only problems involving *conservative forces* (weights and springs) may be solved using *conservation of energy* . Friction or other drag-resistant forces, which depend on velocity or acceleration are nonconservative (a portion of the work done by such forces is transformed into thermal energy which dissipates into the surroundings and may not be recovered). When such forces enter into the problem, use the *principle of work and energy*.

REVIEW QUESTIONS

(Note that much of the material in this Chapter requires a review of Chapter 14)

1. What's the main difference in work and energy methods when they are applied to problems involving rigid bodies as opposed to when they are applied to problems involving particles (Chapter 14)?
2. How would you calculate the work done by a variable external force **F** acting on a rigid body as it moves along the path s?
3. How would you calculate the work done by a couple moment **M** with constant magnitude which causes a rigid body to rotate in the plane through a finite angle θ from θ_1 to θ_2?
4. When is the principle of work and energy used to solve kinetic problems?
5. What is a conservative force? Give some examples of conservative forces.
6. Explain why the weight of a body is a conservative force.
7. Give an example of a nonconservative force and explain why the force is nonconservative.
8. What is potential energy? Give some examples.
9. How can you prove that a force **F** is conservative?
10. When is the conservation of energy equation used to solve problems in kinetics?

19

Planar Kinetics of a Rigid Body:
Impulse and Momentum

MAIN GOALS OF THIS CHAPTER:

- To develop formulations for the linear and angular momentum of a body.
- To apply the principles of linear and angular impulse and momentum to solve rigid-body planar kinetic problems that involve force, velocity, and time.
- To discuss application of the conservation of momentum
- To analyze the mechanics of eccentric impact.

19.1 LINEAR AND ANGULAR MOMENTUM

- **Linear Momentum.** The linear momentum of a rigid body is the vector

$$\mathbf{L} = m\mathbf{v}_G$$

  with magnitude mv_G (units of $kg \cdot m/s$ or $slug \cdot ft/s$) and a direction defined by $\mathbf{v}_G$.

- **Angular Momentum.** The angular momentum of a rigid body is the vector

$$\mathbf{H}_G = I_G\boldsymbol{\omega}$$

  with magnitude $I_G\omega$ (units of $kg \cdot m^2/s$ or $slug \cdot ft^2/s$) and a direction defined by $\boldsymbol{\omega}$, which is always perpendicular to the plane of motion.

  Note The angular momentum of the body can also be computed about a point $P \neq G$. In this case:

$$\mathbf{H}_P = I_G\boldsymbol{\omega} + \text{moment of the linear momentum } m\mathbf{v}G \text{ about } P$$

There are three types of motion to consider:

- **Translation.** When a rigid body is subjected to translation ($\boldsymbol{\omega} = \mathbf{0}$), we obtain

$$L = mv_G$$
$$H_G = 0$$
$$\text{or for some point } A \neq G$$
$$H_A = (d)(mv_G) \;\curvearrowright$$

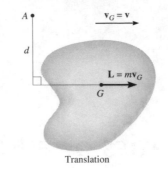

Translation

- **Rotation About a Fixed Axis.** When a rigid body is rotating about a fixed axis passing through point O, we have

$$L = mv_G$$
$$H_G = I_G\omega \quad \text{or} \quad H_O = I_O\omega$$

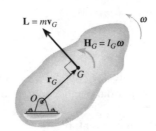

Rotation about a fixed axis

- **General Plane Motion.** When a rigid body is subjected to translation ($\omega = 0$), we obtain

$$L = mv_G$$
$$H_G = I_G\omega$$

or for some point $A \neq G$

$$\overset{+}{\curvearrowright} \quad H_A = I_G\omega + (d)(mv_G)$$

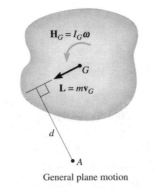

General plane motion

19.2 PRINCIPLE OF IMPULSE AND MOMENTUM

- **Principle of Linear Impulse and Momentum.** The sum of all the impulses created by an *external force system* which acts on the body during the time interval t_1 to t_2 is equal to the change in the linear momentum of the body during the time interval. That is:

$$\sum \int_{t_1}^{t_2} \mathbf{F}\, dt = m\,(\mathbf{v}_G)_2 - m\,(\mathbf{v}_G)_1$$

- **Principle of Angular Impulse and Momentum.** The sum of the angular impulses acting on the body during the time interval t_1 to t_2 is equal to the change in the body's angular momentum during this time interval. That is:

  - **Body in General Plane Motion**

$$\sum \int_{t_1}^{t_2} M_G\, dt = I_G \omega_2 - I_G \omega_1$$

  - **Body Rotating About a Fixed Axis Passing Through** O

$$\sum \int_{t_1}^{t_2} M_O\, dt = I_O \omega_2 - I_O \omega_1$$

- **Motion in the** $x - y$ **Plane.** In the case of planar motion, we have the following three scalar equations:

$$m\,(v + Gx)_1 + \sum \int_{t_1}^{t_2} F_x\, dt = m\,(v_{Gx})_2$$

$$m\,(v_{Gy})_1 + \sum \int_{t_1}^{t_2} F_y\, dt = m\,(v_{Gy})_2 \qquad (19.0)$$

$$I_G \omega_1 + \sum \int_{t_1}^{t_2} M_G\, dt = I_G \omega_2$$

The first two of these equations represent the principle of linear impulse and momentum in the $x - y$ plane and the third equation represents the principle of angular impulse and momentum about the $z - axis$ which passes through the body's mass center G

- **Systems of Connected Bodies.** Equations (19.0) may also be applied to an *entire system of connected bodies* rather than to each body separately. Doing this eliminates the need to include reactive impulses which occur at the connections since they are *internal* to the system.

PROCEDURE FOR SOLVING PROBLEMS

Impulse and momentum principles are used to solve kinetic problems that involve *velocity, force and time,* since these terms are involved in the formulation. We use the following procedure:

- **Free-Body Diagram**

  - Establish the x, y,z inertial frame of reference and draw the free-body diagram in order to account for all the external forces and couple moments that produce impulses on the body.
  - The direction and sense of the initial and final velocity of the body's mass center $\mathbf{v}_G$ and the body's angular velocity ω should be established. If any of these motions is unknown, assume that the sense of its components is in the direction of the positive inertial coordinates.
  - Compute the moment of inertia I_G or I_O.
  - As an alternative procedure, draw the impulse and momentum diagrams for the body or system of bodies. Each of these diagrams represents an outlined shape of the body which graphically accounts for the data required for each of the three terms in Equations (19.0). These diagrams are particularly helpful in visualizing the "moment" terms used in the principle of angular impulse and momentum, if application is about a point other than the body's mass center G or a fixed point O.

- **Principle of Impulse and Momentum**

  - Apply the three scalar equations of impulse and momentum.
  - All the forces acting on the body's free-body diagram will create an impulse; however, some of these forces will do no work.
  - Forces that are functions of time must be integrated to obtain the impulse.
  - The principle of angular impulse and momentum is often used to eliminate unknown impulsive forces that are parallel or pass through a common axis, since the moment of these forces is zero about this axis.

- **Kinematics**

  - If more than three equations are needed for a complete solution, it may be possible to relate the velocity of the body's mass center to the body's angular velocity using *kinematics*.

- See Examples 19-3 to 19-5 in text

19.3 CONSERVATION OF MOMENTUM

- **Conservation of Linear Momentum.** If the sum of all the linear impulses acting on a system of rigid bodies is zero, the linear momentum of the system is constant or *conserved* i.e.,

$$\left(\sum \text{system linear momentum}\right)_1 = \left(\sum \text{system linear momentum}\right)_2$$

  Note It may be possible to apply conservation of linear momentum without inducing appreciable errors even when the linear impulses are small or *nonimpulsive* (small forces acting over very short periods of time e.g., weight of a body).

- **Conservation of Angular Momentum.** The angular momentum of a system of connected rigid bodies is conserved about the system's center of mass G, or a fixed point O, when the sum of all the angular impulses created by the external forces acting on the system is zero or appreciably small (nonimpulsive) when computed about these points i.e.,

$$\left(\sum \text{system angular momentum}\right)_{O1} = \left(\sum \text{system angular momentum}\right)_{O2}$$

  NOTE The angular momentum can be conserved while the linear momentum is not. Such cases occur whenever the external forces creating the linear impulse pass through either the center of mass of the body or a fixed axis of rotation.

PROCEDURE FOR SOLVING PROBLEMS

Provided the initial linear or angular momentum is known, the conservation of linear or angular momentum is used to determine the respective final linear or angular velocity of the body *just after* the time period considered. We use the following procedure:

- **Free-Body Diagram**

  - Establish the x, y inertial frame of reference and draw the free-body diagram for the body or system of bodies during the time of impact. From this diagram, classify each of the applied forces as being wither "impulsive" or "nonimpulsive."
  - By inspection of the free-body diagram, the *conservation of linear momentum* applies in a given direction when no external impulsive forces act on the body or system in that direction; whereas the *conservation of angular momentum* applies about a fixed point O or at the mass center G of a body or system of bodies when all the external forces acting on the body or system create zero moment (or zero angular impulse) about O or G.
  - As an alternative procedure, draw the impulse and momentum diagrams for the body or system of bodies. These diagrams are particularly helpful in visualizing the "moment" terms used in the conservation of angular momentum equation if application is about a point other than the body's mass center G.

- **Conservation of Momentum**

    - Apply the conservation of linear or angular momentum in the appropriate directions.

19.4 ECCENTRIC IMPACT

- *Eccentric Impact* occurs when the line connecting the *mass centers* of the two bodies *does not* coincide with the line of impact. For example:

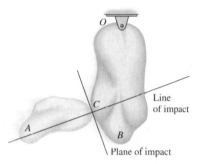

PROCEDURE FOR SOLVING PROBLEMS

- In general, a problem involving the impact of two bodies requires determining the two unknowns $(v_A)_2$ and $(v_B)_2$, assuming $(v_A)_1$ and $(v_B)_1$ are known (or can be determined using kinematics, energy methods, the equations of motion etc.). To solve this problem, two equations must be written:

    - *The first equation generally involves application of the conservation of angular momentum to the two bodies.* In the case of both bodies A and B, we can state that angular momentum is conserved about point O since the impulses at C are internal to the system and the impulses at O create zero moment (or zero angular impulse) about point O.

    - The second equation is obtained using the definition of the coefficient of restitution e :

$$e = \frac{(v_B)_2 - (v_A)_2}{(v_A)_1 - (v_B)_1}.$$

HELPFUL TIPS AND SUGGESTIONS

- Momentum is a *vector* and so has magnitude *and* direction.
- Before applying the equations of impulse and momentum, draw a free-body diagram in order to identify all the forces which cause linear and angular impulses on the body
- If nonimpulsive forces or no impulsive forces act on the body *in a particular direction*, or if the motions of several bodies are involved in the problem, then consider applying the conservation of linear or angular momentum for the solution. Investigation of the free-body diagram (or the impulse diagram) will aid in determining the directions for which the impulsive forces are zero, or axes about which the impulsive forces cause zero angular momentum.

REVIEW QUESTIONS

1. What is meant by *linear momentum*?

2. What is meant by *angular momentum*?

3. True or False? When a body is in translation, the body's linear and angular momentum is given by

$$L = mv_G$$
$$H_A = 0$$

where A is any point on or off the body.

4. True or False? When a body is in general plane motion, the body's linear and angular momentum is given by

$$L = mv_G$$
$$H_A = I_A\omega$$

where A is any point on or off the body.

5. In the case of a system of connected rigid bodies, what is the advantage in applying equations of impulse and momentum to the entire system rather than to each body separately?

6. Explain what is meant by a *nonimpulsive force*.

7. Consider a swimmer who executes a somersault after jumping off a diving board. The swimmer first tucks his arms and legs in close to his chest, then straightens out just before entering the water. Explain why using concepts from impulse and momentum.

8. If angular momentum is conserved is it the case that linear momentum is also always conserved?

20

Three-Dimensional Kinematics
of a Rigid Body

MAIN GOALS OF THIS CHAPTER:

- To analyze the kinematics of a body subjected to rotation about a fixed axis and general plane motion.
- To provide a relative-motion analysis of a rigid body using translating and rotating axes.

20.1 ROTATION ABOUT A FIXED POINT

- **Rotation About a Fixed Point.** In three-dimensions, when a body rotates about a fixed point, the path of motion for a particle P lies on the surface of a sphere having a radius r (distance from the fixed point to P) and centered at the fixed point.

- **Euler's Theorem.** Two "component" rotations about different axes passing through a point are equivalent to a *single resultant rotation* about an axis passing through that point.

- **Finite Rotations.** If the component rotations used in Euler's theorem are *finite* (as opposed to infinitesimal) it is important that the order in which they are applied be maintained. This is because finite rotations *do not obey* the commutative law of addition (e.g., for the two finite rotations θ_1 and θ_2, $\theta_1 + \theta_2 \neq \theta_2 + \theta_1$) and so cannot be classified as vector quantities.

- **Infinitesimal Rotations.** Rotations which are infinitesimally small (e.g., $d\theta$) *may* be classified as vectors since they can be added vectorially in any manner. For this reason, when defining the angular motions of a body subjected to three-dimensional motion, *only rotations which are infinitesimally small* will be considered.

- **Angular Velocity.** If the body is subjected to an angular rotation $d\theta$ about a fixed point, the angular velocity of the body is defined by

$$\boldsymbol{\omega} = \dot{\boldsymbol{\theta}}$$

The line specifying the direction of $\boldsymbol{\omega}$, which is collinear with $d\theta$ is referred to as the *instantaneous axis of rotation*. In general, this axis changes direction during each instant of time. Since $d\theta$ is a vector, so too is $\boldsymbol{\omega}$, and it follows from vector addition that if the body is subjected to two component angular motions, $\boldsymbol{\omega}_1 = \dot{\boldsymbol{\theta}}_1$ and $\boldsymbol{\omega}_2 = \dot{\boldsymbol{\theta}}_2$, the resultant angular velocity is $\boldsymbol{\omega} = \boldsymbol{\omega}_1 + \boldsymbol{\omega}_2$.

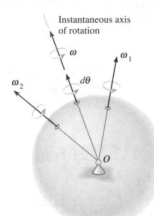

- **Angular Acceleration.** The body's angular acceleration is given by

$$\boldsymbol{\alpha} = \dot{\boldsymbol{\omega}}$$

For motion about a fixed point, $\boldsymbol{\alpha}$ must account for a change in *both* the magnitude and direction of $\boldsymbol{\omega}$, so that, in general, $\boldsymbol{\alpha}$ is *not* directed along the instantaneous axis of rotation.

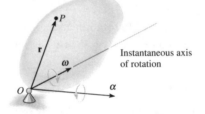

- **Velocity.** Once $\boldsymbol{\omega}$ is specified, the velocity of any point P on a body rotating about a fixed point is given by

$$\mathbf{v} = \boldsymbol{\omega} \times \mathbf{r}$$

where $\mathbf{r}$ defines the position of P measured from the fixed point O.

- **Acceleration.** If $\boldsymbol{\omega}$ and $\boldsymbol{\alpha}$ are known at a given instant, the acceleration of any point P on a body rotating about a fixed point is given by

$$\mathbf{a} = \boldsymbol{\alpha} \times \mathbf{r} + \boldsymbol{\omega} \times (\boldsymbol{\omega} \times \mathbf{r}).$$

Note that the *form* of this equation is the same as that developed in Chapter 16 for planar kinematics.

20.2 THE TIME DERIVATIVE OF A VECTOR MEASURED FROM EITHER A FIXED OR TRANSLATING-ROTATING SYSTEM

-
$$\dot{\mathbf{A}} = (\dot{\mathbf{A}})_{xyz} + \boldsymbol{\Omega} \times \mathbf{A} \tag{20.0}$$
$$\text{where } (\dot{\mathbf{A}})_{xyz} = \dot{A}_x \mathbf{i} + \dot{A}_y \mathbf{j} + \dot{A}_z \mathbf{k}$$

That is, the time derivative of *any vector* $\mathbf{A}$ as observed from the fixed X, Y, Z frame of reference is equal to the time rate of change of $\mathbf{A}$ as observed from the x, y, z translating-rotating frame of reference, $(\dot{\mathbf{A}})_{xyz} = \dot{A}_x \mathbf{i} + \dot{A}_y \mathbf{j} + \dot{A}_z \mathbf{k}$, plus $\boldsymbol{\Omega} \times \mathbf{A}$, the change of $\mathbf{A}$ caused by the rotation of the xyz frame.

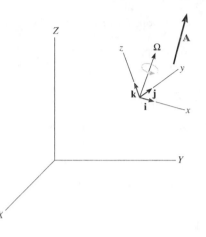

- – Equation (20.0) should always be used whenever $\boldsymbol{\Omega}$ produces a change in the direction of $\mathbf{A}$ as seen from the X, Y, Z reference.
  - – Clearly, if $\boldsymbol{\Omega} = \mathbf{0}$, then $\dot{\mathbf{A}} = \left(\dot{\mathbf{A}}\right)_{xyz}$, and so the time rate of change of $\mathbf{A}$ as observed from both coordinate systems will be the *same*.

20.3 GENERAL MOTION

- Consider a rigid body subjected to general motion in three dimensions for which the angular velocity is $\boldsymbol{\omega}$ and the angular acceleration is $\boldsymbol{\alpha}$. If point A has a known motion of $\mathbf{v}_A$ and $\mathbf{a}_A$, the motion of any other point B is given by

$$\mathbf{v}_B = \mathbf{v}_A + \boldsymbol{\omega} \times \mathbf{r}_{B/A}$$
$$\mathbf{a}_B = \mathbf{a}_A + \boldsymbol{\alpha} \times \mathbf{r}_{B/A} + \boldsymbol{\omega} \times (\boldsymbol{\omega} \times \mathbf{r}_{B/A})$$

These two equations are identical to those describing general *plane* motion of a rigid body (Chapter 16). However, difficulty in application arises for three-dimensional motion, because $\boldsymbol{\alpha}$ measures the change in *both* the direction and magnitude of $\boldsymbol{\omega}$ (in the plane, $\boldsymbol{\alpha}$ measures only a change in the *magnitude* of $\boldsymbol{\omega}$).

20.4 RELATIVE-MOTION ANALYSIS USING TRANSLATING AND ROTATING AXES

- The most general way to analyze the three-dimensional motion of a rigid body requires the use of x, y, z axes that both translate and rotate relative to a second frame X, Y, Z. The main equations are *identical* in form to those used in Section 16.8 for analyzing relative *plane* motion. We review them here briefly:

$$\mathbf{v}_B = \mathbf{v}_A + \boldsymbol{\Omega} \times \mathbf{r}_{B/A} + \left(\mathbf{v}_{B/A}\right)_{xyz} \tag{20.1}$$

$$\mathbf{a}_B = \mathbf{a}_A + \dot{\boldsymbol{\Omega}} \times \mathbf{r}_{B/A} + \boldsymbol{\Omega} \times \left(\boldsymbol{\Omega} \times \mathbf{r}_{B/A}\right) + 2\boldsymbol{\Omega} \times \left(\mathbf{v}_{B/A}\right)_{xyz} + \left(\mathbf{a}_{B/A}\right)_{xyz} \tag{20.2}$$

where

$\mathbf{v}_B, \ \mathbf{a}_B = $ velocity/acceleration of B, measured from the X, Y, Z reference

$\mathbf{v}_A, \ \mathbf{a}_A = $ velocity/acceleration of the origin A of the x, y, z reference, measured from the X, Y, Z reference

$\left(\mathbf{v}_{B/A}\right)_{xyz}, \ \left(\mathbf{a}_{B/A}\right)_{xyz} = $ relative velocity/acceleration of "B with respect to A", as measured by an observer attached to the x, y, z reference

$\boldsymbol{\Omega}, \ \dot{\boldsymbol{\Omega}} = $ angular velocity/acceleration of the x, y, z reference, measured from the X, Y, Z reference.

$\mathbf{r}_{B/A} = $ relative position of "B with respect to A"

The main difference in using these equations *for three-dimensional motion* is that $\dot{\boldsymbol{\Omega}}$ depends on the change in *both* the magnitude and direction of $\boldsymbol{\Omega}$ (in plane motion, $\boldsymbol{\Omega}$ and $\dot{\boldsymbol{\Omega}}$ have a *constant direction* which is always perpendicular to the plane of motion so application of these equations is simplified). Consequently, for three-dimensional motion, $\dot{\boldsymbol{\Omega}}$ must be computed using Equation (20.0) i.e., $\dot{\boldsymbol{\Omega}} = \left(\dot{\boldsymbol{\Omega}}\right)_{xyz} + \boldsymbol{\Omega} \times \mathbf{A}$.

PROCEDURE FOR SOLVING PROBLEMS

Three-dimensional motion of particles or rigid bodies can be analyzed with Equations (20.1) and (20.2) using the following procedure:

- **Coordinate Axes**

  - Select the location and orientation of the X, Y, Z and x, y, z coordinate axes. Most often solutions are easily obtained if at the instant considered

    1. the origins are coincident
    2. the axes are collinear
    3. the axes are parallel

  - If several components of angular velocity are involved in a problem, the calculations will be reduced if the x, y, z axes are selected such that only one component of angular velocity is observed in this frame $\left(\boldsymbol{\Omega}_{xyz}\right)$ and the frame rotates with $\boldsymbol{\Omega}$ defined by the other components of angular velocity.

- **Kinematic Equations**

  - After the origin of the moving reference, A, is defined and the moving point B is specified, Equations (20.1) and (20.2) should be written in symbolic form as

$$\mathbf{v}_B = \mathbf{v}_A + \boldsymbol{\Omega} \times \mathbf{r}_{B/A} + \left(\mathbf{v}_{B/A}\right)_{xyz}$$
$$\mathbf{a}_B = \mathbf{a}_A + \dot{\boldsymbol{\Omega}} \times \mathbf{r}_{B/A} + \boldsymbol{\Omega} \times \left(\boldsymbol{\Omega} \times \mathbf{r}_{B/A}\right) + 2\boldsymbol{\Omega} \times \left(\mathbf{v}_{B/A}\right)_{xyz} + \left(\mathbf{a}_{B/A}\right)_{xyz}$$

  - If $\mathbf{r}_A$ and $\boldsymbol{\Omega}$ appear to *change direction* when observed from the fixed X, Y, Z reference, use a set of primed reference axes x', y', z' having $\boldsymbol{\Omega}' = \boldsymbol{\Omega}$ and Equation (20.0) to determine $\dot{\boldsymbol{\Omega}}$ and the motion $\mathbf{v}_A$ and $\mathbf{a}_A$ of the origin of the moving x, y, z axes.

  - If $\left(\mathbf{r}_{B/A}\right)_{xyz}$ and $\boldsymbol{\Omega}_{xyz}$ appear to *change direction* when observed from the x, y, z reference, use a set of primed reference axes x', y', z' having $\boldsymbol{\Omega}' = \boldsymbol{\Omega}_{xyz}$ and Equation (20.0) to determine $\dot{\boldsymbol{\Omega}}_{xyz}$ and the relative motion $\left(\mathbf{v}_{B/A}\right)_{xyz}$ and $\left(\mathbf{a}_{B/A}\right)_{xyz}$.

  - After the final forms of $\dot{\boldsymbol{\Omega}}$, $\mathbf{v}_A$, $\mathbf{a}_A$, $\dot{\boldsymbol{\Omega}}_{xyz}$, $\left(\mathbf{v}_{B/A}\right)_{xyz}$ and $\left(\mathbf{a}_{B/A}\right)_{xyz}$ are obtained, numerical problem data may be substituted and the kinematic terms evaluated. The components of all these vectors may be selected either along the X, Y, Z axes or along x, y, z. The choice is arbitrary, provided a consistent set of unit vectors is used.

- See Examples 20-4 and 20-5 in text

HELPFUL TIPS AND SUGGESTIONS

- The *vector* form of the kinematic equations for $\mathbf{v}_B$ and $\mathbf{a}_B$ (see Equations (20.1) and (20.2) above and Section 16.8 in text) are identical in both plane and three-dimensional kinematics. This, in itself, is sufficient reason to become comfortable with vector calculus and vector algebra when studying problems in mechanics.

- Examples and worked problems are the key - especially in three-dimensional problems where it is much more difficult to visualize the 'motion.' Work through Examples 20-1 to 20-5 in the text *yourself* before attempting the problems. You will gain a much better understanding of the equations and how to apply them.

REVIEW QUESTIONS

1. True or False? If the component rotations used in Euler's theorem are finite, the order in which they are applied is not important.

2. What is meant by an "infinitesimal rotation"?

3. What's the significance of the angular velocity vector $\boldsymbol{\Omega}$ in Equation (20.0).

4. Can you use Equation (20.0) (and your answer to Question 3) to show that it is meaningless to talk about the angular velocity *of a point* ?

5. Can you prove that the angular velocity vector $\boldsymbol{\Omega}$ satisfying Equation (20.0) is unique? Hint: assume there are two such vectors and show that they necessarily coincide.

6. What's the only difference when using the equations

$$\mathbf{v}_B = \mathbf{v}_A + \boldsymbol{\omega} \times \mathbf{r}_{B/A}$$
$$\mathbf{a}_B = \mathbf{a}_A + \boldsymbol{\alpha} \times \mathbf{r}_{B/A} + \boldsymbol{\omega} \times (\boldsymbol{\omega} \times \mathbf{r}_{B/A})$$

in three-dimensional kinematics as opposed to plane kinematics?

21

Three-Dimensional Kinetics
of a Rigid Body

MAIN GOALS OF THIS CHAPTER:

- To introduce the methods for finding the moments of inertia and products of inertia of a body about various axes.
- To show how to apply the principles of work and energy and linear and angular momentum to a rigid body having three-dimensional motion.
- To develop and apply the equations of motion in three dimensions
- To study the motion of a gyroscope and torque-free motion.

21.1 MOMENTS AND PRODUCTS OF INERTIA

When studying the planar kinetics of a body, it was necessary to introduce the moment of inertia I_G, which was computed about an axis perpendicular to the plane of motion and passing through the body's mass center G. For the kinetic analysis of *three-dimensional motion* it will sometimes be necessary to calculate six inertial quantities. These are called moments and products of inertia and describe, in a particular way, the distribution of mass for a body relative to a given coordinate system that has a specified orientation and point of origin.

- **Moment of Inertia.**

$$I_{xx} = \int_m r_x^2 \, dm = \int_m \left(y^2 + z^2 \right) dm$$

$$I_{yy} = \int_m r_y^2 \, dm = \int_m \left(x^2 + z^2 \right) dm$$

$$I_{zz} = \int_m r_z^2 \, dm = \int_m \left(y^2 + x^2 \right) dm$$

Clearly, the moment of inertia is *always positive*.

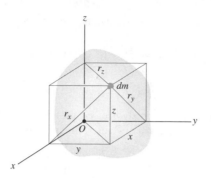

- **Product of Inertia.** The product of inertia is defined with respect to a set of two orthogonal planes as follows:

$$I_{xy} = I_{yx} = \int_m xy\, dm$$

$$I_{yz} = I_{zy} = \int_m yz\, dm$$

$$I_{xz} = I_{zx} = \int_m xz\, dm$$

The product of inertia may be positive, negative or zero. Also, if either one or both of the corresponding orthogonal planes are planes of symmetry for the mass, the product of inertia with respect to these planes will be zero.

- **Parallel-Axis Theorem.** If the body's mass center G has coordinates x_G, y_G, z_G then the parallel-axis equations used to calculate the moments of inertia about the x, y, z axes are

$$I_{xx} = (I_{x'x'})_G + m\left(y_G^2 + z_G^2\right)$$
$$I_{yy} = (I_{y'y'})_G + m\left(x_G^2 + z_G^2\right)$$
$$I_{zz} = (I_{z'z'})_G + m\left(x_G^2 + y_G^2\right)$$

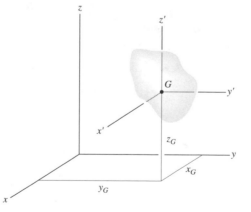

- **Parallel-Plane Theorem.** The parallel-plane theorem is used to transfer the products of inertia of the body from a set of three orthogonal planes passing through the body's mass center to a corresponding set of three parallel planes passing through some other point O If x_G, y_G, z_G are the perpendicular distances between the planes, the parallel-plane equations are:

$$I_{xy} = \left(I_{x'y'}\right)_G + m x_G y_G$$
$$I_{yz} = \left(I_{y'z'}\right)_G + m y_G z_G$$
$$I_{zx} = \left(I_{z'x'}\right)_G + m z_G x_G$$

• **Inertia Tensor**. The inertial properties of a body are completely characterized by nine terms, six of which are independent. We organize these terms in an array called the *inertial tensor*:

$$\begin{pmatrix} I_{xx} & -I_{xy} & -I_{xz} \\ -I_{yx} & I_{yy} & -I_{yz} \\ -I_{zx} & -I_{zy} & I_{zz} \end{pmatrix}$$

The inertial tensor has a unique set of values for a body when it is computed for each location of the origin O and orientation of the coordinate axes.

• **Principal Moments and Axes of Inertia.** In general, for point O, we can specify a unique axes inclination for which the products of inertia for the body are zero when computed with respect to these axes. When this is done, the inertia tensor is *diagonalized* i.e., it can be written as

$$\begin{pmatrix} I_x & 0 & 0 \\ 0 & I_y & 0 \\ 0 & 0 & I_z \end{pmatrix}$$

where $I_x = I_{xx}$, $I_y = I_{yy}$ and $I_z = I_{zz}$ are termed *principal moments of inertia* for the body which are computed from the *principal axes of inertia*. Of the three principal moments of inertia, one will be a maximum and another a minimum of the body's moment of inertia.

• **Moment of Inertia About an Arbitrary Axis.** Suppose we have calculated all 9 components of the inertia tensor for the x, y, z axes having an origin at O and we now wish to determine the moment of inertia of the body about the Oa axis, whose direction is defined by the unit vector $\mathbf{u}_a$.

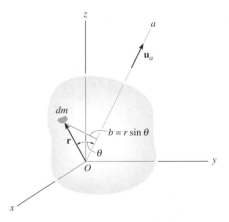

We have

$$I_{Oa} = I_{xx} u_x^2 + I_{yy} u_y^2 + I_{zz} u_z^2 - 2 I_{xy} u_x u_y - 2 I_{yz} u_y u_z - 2 I_{zx} u_z u_x$$

Here, $\mathbf{u}_a = u_x \mathbf{i} + u_y \mathbf{j} + u_z \mathbf{k}$.

21.2 ANGULAR MOMENTUM

We develop necessary equations used to determine the angular momentum of a rigid body about an arbitrary point. This will allow us to again use the principal of impulse and momentum to solve kinetic problems (in three-dimensions) involving force, velocity and time:

- **About a Fixed Point** O

$$\mathbf{H}_O = \int_m \boldsymbol{\rho}_O \times (\boldsymbol{\omega} \times \boldsymbol{\rho}_O)\, dm$$

- **About Center of Mass** G

$$\mathbf{H}_G = \int_m \boldsymbol{\rho}_G \times (\boldsymbol{\omega} \times \boldsymbol{\rho}_G)\, dm$$

- **About Arbitrary Point** A

$$\mathbf{H}_A = \boldsymbol{\rho}_{G/A} \times m\mathbf{v}_G + \mathbf{H}_G$$

- **Rectangular Components of H.** In scalar component form, $\mathbf{H}_O$ or $\mathbf{H}_G$ is given by

$$H_x = I_{xx}\omega_x - I_{xy}\omega_y - I_{xz}\omega_z$$
$$H_y = -I_{yx}\omega_x + I_{yy}\omega_y - I_{yz}\omega_z$$
$$H_z = -I_{zx}\omega_x - I_{zy}\omega_y + I_{zz}\omega_z$$

Note If the x, y, z coordinate axes are oriented such that they become *principal axes of inertia* for the body at the point, the three components of angular momentum become:

$$H_x = I_x\omega_x$$
$$H_y = I_y\omega_y$$
$$H_z = I_z\omega_z$$

- **Principle of Impulse and Momentum.** We can now use the principal of impulse and momentum to again solve kinetic problems involving *force, velocity and time*:

$$m\,(\mathbf{v}_G)_1 + \sum \int_{t_1}^{t_2} \mathbf{F}\, dt = m\,(\mathbf{v}_G)_2$$
$$(\mathbf{H}_O)_1 + \sum \int_{t_1}^{t_2} \mathbf{M}_O\, dt = (\mathbf{H}_O)_2$$

These are *six* scalar equations (remember that in three-dimensions each vector has *three* components).

21.3 KINETIC ENERGY

We develop necessary equations used to determine the *kinetic energy* of a rigid body. This will allow us to again use the principal of work and energy to solve kinetic problems (in three-dimensions) involving force, velocity and displacement. The kinetic energy of the ith particle of the body having a mass m_i and velocity $\mathbf{v}_i$, measured relative to an inertial X, Y, Z frame of reference is

$$T_i = \frac{1}{2}m_i v_i^2 = \frac{1}{2}(\mathbf{v}_i \cdot \mathbf{v}_i)$$

Provided the velocity of an arbitrary point A in the body is known, we can relate $\mathbf{v}_i$ to $\mathbf{v}_A$. The following special cases arise:

- **A is a Fixed Point O.**

$$T = \frac{1}{2}\boldsymbol{\omega} \cdot \mathbf{H}_O$$

If the x, y, z axes are principal axes of inertia then:

$$T = \frac{1}{2}\left(Ix\omega_x^2 + Iy\omega_y^2 + Iz\omega_z^2\right)$$

- **A is the Center of Mass G.**

$$T = \frac{1}{2}mv_G^2 + \frac{1}{2}\boldsymbol{\omega} \cdot \mathbf{H}_G$$

If the x, y, z axes are principal axes of inertia then:

$$T = \frac{1}{2}mvG^2 + \frac{1}{2}\left(I_x\omega_x^2 + I_y\omega_y^2 + I_z\omega_z^2\right)$$

Note that the kinetic energy consists of two parts: the translational kinetic energy of the mass center $\frac{1}{2}mv_G^2$ and the body's rotational kinetic energy.

- **Principle of Work and Energy.** We can now use the principal of work and energy to again solve kinetic problems involving *force, velocity and displacement*. For each body, we have:

$$T_1 + \sum U_{1-2} = T_2$$

This is *only one scalar equation* (of the same form as planar kinetics).

21.4 EQUATIONS OF MOTION

- **Equations of Translational Motion.**

$$\sum \mathbf{F} = m\mathbf{a}_G$$

or, as three scalar equations

$$\sum F_x = m\,(a_G)_x$$
$$\sum F_y = m\,(a_G)_y$$
$$\sum F_z = m\,(a_G)_z$$

Here, $\sum \mathbf{F} = \sum F_x\mathbf{i} + \sum F_y\mathbf{j} + \sum F_z\mathbf{k}$ is the sum of all the external forces acting on the body and x, y, z is an inertial frame..

- **Equations of Rotational Motion.** Consider the inertial frame X, Y, Z and a set of axes x, y, z with origin at G, the body's mass center. In general, we assume that the x, y, z axes are rotating with an angular velocity $\boldsymbol{\Omega}$ and that the body rotates with an angular velocity $\boldsymbol{\omega}$. Depending on whether we sum moments about G or a fixed point O, we have the following two forms of the rotational equation of motion ($\sum \mathbf{M}_O = \dot{\mathbf{H}}_O$ or $\sum \mathbf{M}_G = \dot{\mathbf{H}}_G$):

$$\sum \mathbf{M}_O = \left(\dot{\mathbf{H}}_O\right)_{xyz} + \boldsymbol{\Omega} \times \mathbf{H}_O$$
$$\sum \mathbf{M}_G = \left(\dot{\mathbf{H}}_G\right)_{xyz} + \boldsymbol{\Omega} \times \mathbf{H}_G$$

Here, $\sum \mathbf{M}$ is the sum of the moments of all external forces acting on the rigid body about point O or G, $\left(\dot{\mathbf{H}}\right)_{xyz}$ is the time rate of change of $\mathbf{H}$ measured from the x, y, z reference.

There are three ways in which one can define the motion of the x, y, z axes. Obviously, motion of the x, y, z axes should be *chosen* to yield the simplest set of moment equations for the solution of a particular problem.

1. x, y, z **Axes Having Motion** $\boldsymbol{\Omega} = 0$. If the body has general motion, the x, y, z axes may be chosen with origin at G, such that the axes only translate relative to the inertial X, Y, Z frame. At first glance this seems like a simplification since $\boldsymbol{\Omega} = 0$ but the body may have a rotation $\boldsymbol{\omega}$ about these axes and therefore the moments and products of inertia of the body would have to be expressed as functions of time - not easy! So this choice of axes x, y, z has little value.

2. x, y, z **Axes Having Motion** $\boldsymbol{\Omega} = \boldsymbol{\omega}$. The x, y, z axes may be chosen such that they are *fixed in and move* with the body. The moments and products of inertia relative to these axes will be *constant* during the motion. The rotational equations of motion become

$$\sum \mathbf{M}_O = \left(\dot{\mathbf{H}}_O\right)_{xyz} + \boldsymbol{\omega} \times \mathbf{H}_O$$
$$\sum \mathbf{M}_G = \left(\dot{\mathbf{H}}_G\right)_{xyz} + \boldsymbol{\omega} \times \mathbf{H}_G$$

Further, if the x, y, z axes are chosen as *principal axes* of inertia, we obtain the rotational equations:

$$\sum M_x = I_x \dot{\omega}_x - \left(I_y - I_z\right) \omega_y \omega_z$$
$$\sum M_y = I_y \dot{\omega}_y - \left(I_z - I_x\right) \omega_z \omega_x \qquad (21.0)$$
$$\sum M_z = I_z \dot{\omega}_z - \left(I_x - I_y\right) \omega_x \omega_y$$

These are known as the Euler equations of motion. They apply *only for moments summed about a fixed point O or about the body's mass center G.*

3. x, y, z **Axes Having Motion** $\boldsymbol{\Omega} \neq \boldsymbol{\omega}$. This case is particularly suitable for the analysis of spinning tops and gyroscopes which are symmetrical about their spinning axes. In this case, the moments and products of inertia relative to these axes remain *constant* during the motion. Further, if the x, y, z axes are chosen as principal axes of inertia, we obtain the rotational equations:

$$\sum M_x = I_x \dot{\omega}_x - I_y \Omega_z \omega_y + I_z \Omega_y \omega_z$$
$$\sum M_y = I_y \dot{\omega}_y - I_z \Omega_x \omega_z + I_x \Omega_z \omega_x$$
$$\sum M_z = I_z \dot{\omega}_z - I_x \Omega_y \omega_x + I_y \Omega_x \omega_y$$

Here, Ω_x, Ω_y, Ω_z represent the x, y, z components of $\boldsymbol{\Omega}$ measured from the inertial frame of reference and $\dot{\omega}_x$, $\dot{\omega}_y$, $\dot{\omega}_z$ must be determined relative to the x, y, z axes that have the rotation $\boldsymbol{\Omega}$ (see Example 21-6 in text).

Unfortunately, both sets of rotational equations obtained thus far are a series of *three-first-order coupled nonlinear differential equations*. This makes them extremely difficult to solve in general. In fact, success in solving either set of equations depends greatly on what is given and what is to be solved for. Consequently, numerical methods are often used to obtain solutions.

PROCEDURE FOR SOLVING PROBLEMS

Problems involving the three-dimensional motion of a rigid body can be solved using the following procedure:

- **Free-Body Diagram**

  - Draw a free-body diagram of the body at the instant considered and specify the x, y, z coordinate system. The origin of this reference must be located either at the body's mass center G, or at the point O, considered fixed in an inertial frame and located either in the body or on a massless extension of the body.

  - Unknown reactive forces can be shown having a positive sense of direction.

  - Depending on the nature of the problem, decide what type of rotational motion $\boldsymbol{\Omega}$ the x, y, z coordinate system should have i.e., $\boldsymbol{\Omega} = 0$, $\boldsymbol{\Omega} = \boldsymbol{\omega}$, $\boldsymbol{\Omega} \neq \boldsymbol{\omega}$. When choosing, one should keep in mind that the moment equations are simplified when the axes move in such a manner that they represent principal axes of inertia for the body at all times.

  - Compute the necessary moments and products of inertia for the body relative to the x, y, z axes.

- **Kinematics**

  - Determine the x, y, z components of the body's angular velocity and compute the time derivatives of $\boldsymbol{\omega}$.

  - Note that if $\boldsymbol{\Omega} = \boldsymbol{\omega}$ then $\dot{\boldsymbol{\omega}} = (\dot{\boldsymbol{\omega}})_{xyz}$, and we can either find the components of $\boldsymbol{\omega}$ along the x, y, z axes when the axes are oriented in a general position, and then take the time derivative of the magnitudes of these components, $(\dot{\boldsymbol{\omega}})_{xyz}$; or we can find the time derivative of $\boldsymbol{\omega}$ with respect the X, Y, Z axes $\dot{\boldsymbol{\omega}}$, and then determine the components $\dot{\omega}_x, \dot{\omega}_y, \dot{\omega}_z$.

- **Equations of Motion**

  - Apply either the two vector equations

  $$\sum \mathbf{F} = m\mathbf{a}_G$$

  and

  $$\sum \mathbf{M}_O = \left(\dot{\mathbf{H}}_O\right)_{xyz} + \boldsymbol{\Omega} \times \mathbf{H}_O$$

  or

  $$\sum \mathbf{M}_G = \left(\dot{\mathbf{H}}_G\right)_{xyz} + \boldsymbol{\Omega} \times \mathbf{H}_G$$

  or the six scalar component equations (3 of translation and 3 of rotation) appropriate for the x, y, z coordinate axes chosen for the problem.

  ∗ **Examples.** See Examples 21-4 to 21-6 in text.

21.5 GYROSCOPIC MOTION

In this section we develop the equations defining the motion of a body (top) which is symmetrical with respect to an axis and moving about a fixed point lying on the axis. The body's motion is analyzed using *Euler angles* ϕ, θ, ψ. The angular velocity components $\dot{\phi}, \dot{\theta}$ and $\dot{\psi}$ are known as the *precession, nutation, and spin*, respectively.

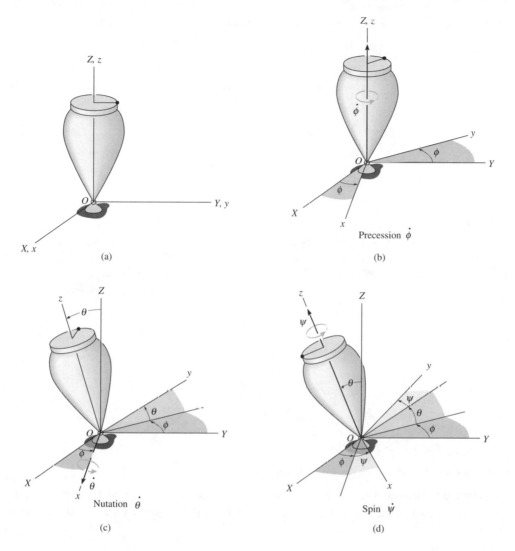

(a)

(b)

Precession $\dot{\phi}$

(c)

Nutation $\dot{\theta}$

(d)

Spin $\dot{\psi}$

• **Rotational Equations of Motion**

$$\sum M_x = I\left(\ddot{\theta} - \dot{\phi}^2 \sin\theta \cos\theta\right) + I_z \dot{\phi} \sin\theta \left(\dot{\phi} \cos\theta + \dot{\psi}\right)$$

$$\sum M_y = I\left(\ddot{\phi} \sin\theta + 2\dot{\phi}\dot{\theta} \cos\theta\right) - I_z \dot{\theta} \left(\dot{\phi} \cos\theta + \dot{\psi}\right)$$

$$\sum M_z = I_z \left(\ddot{\psi} + \ddot{\phi} \cos\theta - \dot{\phi}\dot{\theta} \sin\theta\right)$$

where x, y, z axes represent principal axes of inertia of the body for any spin of the body about these axes. Hence the moments of inertia are constant and we write $I = I_{xx} = I_{yy}$ and $I_z = I_{zz}$. Each moment summation applies only at the fixed point O or at G. These equations are a coupled set of nonlinear second order differential equations. Consequently, it may not be possible to obtain a closed-form solution. There is a *special case*, however.

    – **Steady Precession ($\theta, \dot{\phi}, \dot{\psi}$ all constant).** The rotational equations now become

$$\sum M_x = -I\dot{\phi}^2 \sin\theta \cos\theta + I_z\dot{\phi}\sin\theta \left(\dot{\phi}\cos\theta + \dot{\psi}\right)$$

or

$$\sum M_x = \dot{\phi}\sin\theta(I_z\omega_z - I\dot{\phi}\cos\theta)$$

and

$$\sum M_y = 0$$
$$\sum M_z = 0$$

- **Examples.** See Examples 21-7 and 21-8 in text.

21.6 TORQUE-FREE MOTION

When the only external force acting on a body is caused by gravitation, the general motion of the body is referred to as torque-free motion. This type of motion is characteristic of planets, artificial satellites and projectiles - provided the effects of air friction are neglected. We have the following results describing the motion:

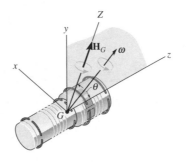

- **For torque-free motion of an axisymmetrical body:**

  1. The angle θ formed between the angular-momentum vector and the spin of the body remains constant.
  2. The angular momentum $\mathbf{H}_G$, precession $\dot{\phi}$ and spin $\dot{\psi}$ for the body remain constant at all times during the motion.

- **Functional Equations**

$$\boldsymbol{\omega} = \frac{H_G \sin\theta}{I}\mathbf{j} + \frac{H_G \cos\theta}{I_z}\mathbf{k}$$

$$\theta = const.$$

$$\dot{\phi} = \frac{H_G}{I}$$

$$\dot{\psi} = \frac{I - I_z}{I I_z} H_G \cos\theta$$

or

$$\dot{\psi} = \frac{I - I_z}{I_z}\dot{\phi}\cos\theta$$

   Here, the origin of the x, y, z system is located at G such that $I_{zz} = I_z$, and $I_{xx} = I_{yy} = I$ for the body and $\boldsymbol{\omega}$ represents the body's angular velocity.

- **Examples.** See Example 21-9 in text.

HELPFUL TIPS AND SUGGESTIONS

- Examples and worked problems are the key - especially in three-dimensional problems where it is much more difficult to visualize the 'motion.' Work through Examples 21-1 to 21-9 in the text *yourself* before attempting the problems. You will gain a much better understanding of the equations and how to apply them.

- Be careful not to 'get lost' in the detail. Make a concise list of the relevant equations (you might ask your professor to identify these e.g., for exam review purposes) and practice with them.

REVIEW QUESTIONS

1. True or False? General three-dimensional motion is a much more difficult subject than plane motion mainly because neither the kinematics nor kinetics differential equations governing the rotational motion of the body are linear.

2. How many different moments of inertia are there in three-dimensions? What about products of inertia?

3. What's the inertia tensor? What's it used for?

4. In what variables are the Euler equations (21.0) nonlinear?

5. Does the plane motion equation $\sum M_G = I_G \omega$ extend simply to general three-dimensional motion?

6. What's the main difference between the principle of work and energy in three-dimensions as opposed to in the plane?

22

Vibrations

MAIN GOALS OF THIS CHAPTER:

- To discuss undamped one-degree-of-freedom vibration of a rigid body using the equation of motion and energy methods.
- To study the analysis of undamped forced vibration and viscous damped forced vibration.
- To introduce the concept of electrical circuit analogs to study vibrational motion.

22.1 UNDAMPED FREE VIBRATION

- A *vibration* is the periodic motion of a body or system of connected bodies displaced from a position of equilibrium.

    – *Free vibration* - when the motion is maintained by gravitational or elastic restoring forces.
    – *Forced vibration* - caused by an external periodic or intermittent force applied to the system

- *Undamped vibrations* can continue indefinitely since frictional effects are neglected in the analysis.
- *Damped vibrations* are vibrations which decay/die out with time due to the effects of internal and external frictional forces. In reality, the motion of all vibrating bodies is *actually damped*.

Equation for Undamped Free Vibration of a One-Degree-of-Freedom System

If a vibrating system has a *single degree-of-freedom*, it requires only *one coordinate* to specify completely the position of the system at any time t.

- **Equation for Undamped Free Vibration of a Simple Block-Spring System.** One of the simplest examples of a single degree-of-freedom system for undamped free vibrations is the block-spring system shown below.

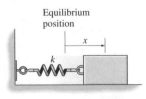

In this case, the block's motion may be determined by applying the equation of motion to the block when it is in the displaced position x. We obtain the following equation describing the block's vibrations (*simple harmonic motion - acceleration is proportional to the block's displacement*):

$$\ddot{x} + p^2 x = 0, \quad p = \sqrt{\frac{k}{m}}$$

Here, p is called the *circular frequency (rad/s)*, x is the displacement from equilibrium, m is the mass of the vibrating block attached to a spring of stiffness k.

The general solution of this differential equation is

$$x(t) = A \sin pt + B \cos pt$$

where A and B are two constants of integration generally determined from the initial conditions of the problem. An alternative form of this general solution is

$$x(t) = C \sin (pt + \phi)$$

where again C and ϕ are new arbitrary constants to be determined. We have the following vibrating characteristics of the block:

- **Amplitude**: $C = \sqrt{A^2 + B^2}$ is maximum displacement of the body
- **Period**: $\tau = \frac{2\pi}{p}$ is time required to complete one cycle
- **Frequency**: The frequency is the number of cycles completed per unit of time. It is the reciprocal of the period:

$$f = \frac{1}{\tau} = \frac{p}{2\pi} = \frac{1}{2\pi}\sqrt{\frac{k}{m}}$$

The frequency is expressed in cycles/s i.e., $1Hz$ (Hertz) $= 1cycle/s = 2\pi rad/s$

- **Natural Frequency and Other Single Degree-of-Freedom Systems.** When a body or system of connected bodies is given an initial displacement from its equilibrium position and released, it will vibrate with a definite frequency known as the *natural frequency*. Provided the body has a single degree of freedom, that is, it requires only one coordinate to specify completely the position of the system at any time, then the vibrating motion of the body will have the *same characteristics* as the simple harmonic motion of the block and spring just mentioned above. Consequently, the body's motion is described by a differential equation of the same "standard form" i.e.,

$$\ddot{x} + p^2 x = 0 \tag{22.0}$$

Hence, if the circular frequency p of the body is known, the period of vibration τ, natural frequency f and other vibrating characteristics of the body can be established using the same equations as those developed for the block and spring:

$$x(t) = A \sin pt + B \cos pt$$
$$x(t) = C \sin (pt + \phi)$$
$$C = \sqrt{A^2 + B^2} \text{ is the } amplitude$$
$$\tau = \frac{2\pi}{p} \text{ is the } period$$
$$f = \frac{1}{\tau} \text{ is the } natural\ frequency$$

PROCEDURE FOR SOLVING PROBLEMS

As in the case of the block and spring, the circular frequency p of a rigid body or system of connected rigid bodies having a single degree of freedom can be determined as follows:

- **Free-Body Diagram**

    - Draw a free-body diagram of the body when the body is displaced by a *small amount* from its equilibrium position.
    - Locate the body with respect to its equilibrium position by using an appropriate inertial coordinate e.g., q. The acceleration of the body's mass center $\mathbf{a}_G$ or the body's angular acceleration α should have a sense which is in the positive direction of the position coordinate.
    - If the rotational equation of motion $\sum M_P = \sum (M_k)_P$ is to be used, then it may be beneficial to also draw the kinetic diagram since it graphically accounts for the components $m\,(\mathbf{a}_G)_x$, $m\,(\mathbf{a}_G)_y$ and $I_G\alpha$, and thereby makes it convenient for visualizing the terms needed in the moment sum $\sum (\mathcal{M}_k)_P$.

- **Equation of Motion**

    - Apply the equation of motion to relate the elastic or gravitational *restoring* forces and couple moments acting on the body to the body's accelerated motion.

- **Kinematics**

    - Using kinematics, express the body's accelerated motion in terms of the second time derivative of the position coordinate.
    - Substitute the result into the equation of motion and determine p by rearranging the terms so that the resulting equation is of the "standard form" (22.0) i.e., $\ddot{q} + p^2 q = 0$..

- **Examples.** See Examples 22-1 to 22-4 in text.

22.2 ENERGY METHODS

The simple harmonic motion of a body is due only to gravitational and elastic restoring forces acting on the body. Since these types of forces are conservative, it is also possible to use the conservation of energy equation to obtain the body's natural frequency or period of vibration. We use the following procedure:

PROCEDURE FOR SOLVING PROBLEMS

- **Energy Equation**

    - Draw the body when it is displaced by a *small amount* from its equilibrium position and define the location of the body from its equilibrium position by an appropriate position coordinate q.
    - Formulate the equation of energy for the body, $T + V = $ constant, in terms of the position coordinate.
    - In general, the kinetic energy must account for both the body's translational and rotational motion, $T = \frac{1}{2}mv_G^2 + \frac{1}{2}I_G\omega^2$.
    - The potential energy is the sum of the gravitational and elastic potential energies of the body, $V = V_g + V_e$. In particular, V_g should be measured from a datum for which $q = 0$ (equilibrium position).

- **Time Derivative**

    - Take the time derivative of the energy equation using the chain rule of calculus and factor out the common terms. The resultant differential equation represents the equation of motion for the system. The value of p is obtained after rearranging the terms in the "standard form" $\ddot{q} + p^2 q = 0$.

Note. To obtain the "standard form" $\ddot{q} + p^2q = 0$, it may be necessary to approximate e.g., $\sin q$ by the first term in its power series expansion i.e., $\sin q \sim q$ and $\cos q$ by the first two terms of its power series expansion e.g., $\cos q \sim 1 - \frac{q^2}{2}$. Such approximations are justified in that we are interested only in *small vibrations* (small angles etc).

- **Examples.** See Examples 22-5 to 22-6 in text.

22.3 UNDAMPED FORCED VIBRATION

Undamped forced vibration is considered to be one of the most important types of vibrating motion in engineering work.

- **Periodic Force.** The block and spring system again provides a convenient model for representing the vibrational characteristics of a system subjected to a periodic force $F = F_0 \sin \omega t$. This force has an amplitude of F_0 and a forcing frequency ω.

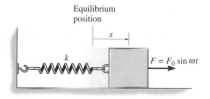

The corresponding equation describing vibrations of the block is obtained on application to the equation of motion and is given by

$$\ddot{x} + p^2 x = \frac{F_0}{m} \sin \omega t, \quad p = \sqrt{\frac{k}{m}} \tag{22.1}$$

The general solution of this equation is given by

$$x(t) = x_c + x_p$$

$$= A \sin pt + B \cos pt + \underbrace{\frac{\frac{F_0}{k}}{1 - \left(\frac{\omega}{p}\right)^2} \sin \omega t}$$

$\underbrace{}$
x_c describes free vibrations

x_p describes forced vibration caused by applied force

Note If the force is applied with a frequency close to the natural frequency of the system i.e., $\omega \approx p$, the amplitude of vibration of the block becomes extremely large. This condition is called *resonance*, and, in practice, resonating vibrations can cause tremendous stress and rapid failure of parts.

- **Periodic Support Displacement.** Forced vibrations can also arise from the periodic excitation of the support of a system. In this case, the governing equation is identical to (22.1) with F_0 replaced by $k\delta_0$ where δ_0 is defined by the support displacement $\delta = \delta_0 \sin \omega t$

22.4 VISCOUS DAMPED FREE VIBRATION

Since real vibrations die out with time, the presence of damping forces should be included in the vibration analysis. In many cases, damping is attributed to the resistance created by the substance (e.g., water, oil or air) in which the system vibrates.

- **Viscous Damping Force.** Provided the body moves slowly through the substance in which the body vibrates, the magnitude F of the viscous damping force is given by

$$F = c\dot{x}$$

where the constant c is called the coefficient of viscous damping and has units of $N \cdot s/m$ or $lb \cdot s/ft$.

- **Equation of Viscous Damped Free Vibration.** The vibrating motion of a body or system having viscous damping may again be characterized by a block-spring system in which damping effects (dashpot) are included. The equation describing subsequent vibrations of the block is given by

$$m\ddot{x} + c\dot{x} + kx = 0 \tag{22.2}$$

  - **General Solution.** The general solution of this differential equation takes different forms depending on the two values

$$\lambda_1 = -\frac{c}{2m} + \sqrt{\left(\frac{c}{2m}\right)^2 - \frac{k}{m}} \tag{22.3}$$

$$\lambda_2 = -\frac{c}{2m} - \sqrt{\left(\frac{c}{2m}\right)^2 - \frac{k}{m}} \tag{22.4}$$

Clearly the nature of λ_1 and λ_2 will depend on the value of c which makes the radical equal to zero. That is, $c = c_c = 2m\sqrt{\frac{k}{m}} = 2mp$. This is called the *critical damping coefficient* c_c.

- **Overdamped System.** When $c > c_c$, λ_1 and λ_2 are both real and the general solution of (22.2) takes the form

$$x(t) = Ae^{\lambda_1 t} + Be^{\lambda_2 t} \tag{22.5}$$

where A, B are constants to be determined from initial conditions. Motion corresponding to this solution is *nonvibrating* - damping is so strong that the block simply creeps back to its original position without oscillating. The system is said to be *overdamped*.

- **Critically Damped System.** If $c = c_c$ then $\lambda_1 = \lambda_2 = -\frac{c_c}{2m} = -p$. This is known as *critical damping*, since it represents a condition where c has the smallest value necessary to cause the system to be nonvibrating. The general solution of (22.2) in this case is given by

$$x(t) = (A + Bt)e^{-pt} \tag{22.6}$$

- **Underdamped System.** Most often, $c < c_c$, in which case the system is underdamped. Then, λ_1 and λ_2 are both complex The general solution of (22.2) in this case is given by

$$x(t) = D\left[e^{-\left(\frac{c}{2m}\right)t} \sin(p_d t + \phi)\right] \tag{22.7}$$

where D and ϕ are constants generally determined from the initial conditions. The constant p_d is called the *damped natural frequency of the system*:

$$p_d = \sqrt{\frac{k}{m} - \left(\frac{c}{2m}\right)^2}$$

$$= p\sqrt{1 - \left(\frac{c}{c_c}\right)^2}$$

where the ratio $\frac{c}{c_c}$ is called the *damping factor*.

Note The period of damped vibration can be written as

$$\tau_d = \frac{2\pi}{p_d}$$

Since $p_d < p$, $\tau_d > \tau = \frac{2\pi}{p}$, the period of free vibration.

22.5 VISCOUS DAMPED FORCED VIBRATION

The most general case of single-degree-of-freedom vibrating motion occurs when the system includes the effects of forced motion *and* induced damping. The differential equation which describes the motion of a block and spring system which includes the effects of forced motion *and* induced damping is:

$$m\ddot{x} + c\dot{x} + kx = F_0 \sin \omega t \tag{22.8}$$

Note that a similar equation is obtained for a block and spring having a periodic support displacement, which includes the effects of damping. In this case, F_0 in (22.8) is simply replaced by $k\delta_0$

- **General Solution of Equation (22.5)**. The general solution of Equation (22.5) is of the form

$$x(t) = x_c(t) + x_p(t)$$

where $x_c(t)$ is the general solution of Equation (22.2) (which depends on the values of λ_1 and λ_2 in (22.3) and (22.4) - see (22.5)–(22.7)) and

$$x_p(t) = C' \sin(\omega t - \phi\prime)$$

where the constants C' and ϕ' are given by

$$C' = \frac{\frac{F_0}{k}}{\sqrt{\left[1 - \left(\frac{\omega}{p}\right)^2\right]^2 + \left[2\left(\frac{c}{c_c}\right)\left(\frac{\omega}{p}\right)\right]^2}},$$

$$\phi' = \tan^{-1}\left[\frac{2\left(\frac{c}{c_c}\right)\left(\frac{\omega}{p}\right)}{1 - \left(\frac{\omega}{p}\right)^2}\right]$$

The angle ϕ' represents the phase difference between the applied force and the resulting steady-state vibration of the damped system.

HELPFUL TIPS AND SUGGESTIONS

- When dealing with undamped free vibration, the differential equation describing vibrations always takes the standard form (22.0). This provides a useful way of 'checking' your work when analyzing vibrations of a system other than the block-spring system (see Examples 22-1 to 22-4).

- It may be useful to review some elementary techniques for solving linear ordinary differential equations with constant coefficients.

REVIEW QUESTIONS

1. When does free vibration occur?
2. Write down the differential equation describing undamped free vibrations, its general solution and the definitions of 'amplitude,' 'period,' and 'frequency.'
3. What is the 'natural frequency' of a vibrating system?
4. What is 'simple harmonic motion'?
5. Why is it possible to use conservation of energy to obtain the natural frequency of a body executing undamped free vibrations.?
6. What is resonance and how does it occur?
7. In damped free vibrations of a system, when is it the case that vibrations die out in time?
8. Write down the equation describing damped forced vibrations of a block-spring system. Describe the solution.

ANSWERS TO REVIEW QUESTIONS

Chapter 12:

1. Yes because a particle is modelled as a point (i.e., an object with mass but negligible size/shape).

2. Yes.

3. No. For example, a particle travelling in a general curvilinear path has both normal and tangential acceleration components (see Section 12.7).

4. No. Since the ball has constant speed, from Eqs. (12.3)–(12.6), the tangential component of acceleration is always zero **but** the normal component of acceleration is non-zero. This makes the acceleration of the center of the ball non-zero.

5. No. The velocity depends on the reference frame. For example, we can always define a reference frame with respect to which the particle isn't moving (i.e., the particle stays fixed for all time in that reference frame), in which case, the particle's velocity (and acceleration) will always be zero.

6. This happens (see Equation 12.6) when $a_n = \frac{v^2}{\rho} \equiv 0 \Longleftrightarrow v = 0$ or $\rho \to \infty$ i.e., when the particle is fixed for all time or when the radius of curvature is infinite i.e., when the particle is moving in a straight line (rectilinear motion).

7. No. For example, if the rectilinear motion of a point P is described by $x(t) = t^2$, then

$$\text{speed:} \quad \dot{x}(t) = 2t$$
$$\text{acceleration:} \quad \ddot{x}(t) = 2$$

   Clearly the speed is zero at $t = 0$ but the acceleration is constant $(= 2)$ for all time.

8. From Eq. (12.8), the radial component of acceleration for general curvilinear motion is given by $a_r = \ddot{r} - r\dot{\theta}^2$. If $\ddot{r} = 0$, $a_r = -r\dot{\theta}^2 \neq 0$.

Chapter 13

1. **(a)** True.

   (b) False. The equation of motion is based on experimental evidence.

2. See Equations (13.3):

$$\sum F_x = ma_x,$$
$$\sum F_y = ma_y.$$

3. When a problem involves the motion of a particle along a known curved path - since the acceleration components can be readily formulated - see Section 13.5.

4. Since the particle is constrained to move *along* the path.

5. When data regarding the angular motion of the radial line r are given, or in cases where the path can be conveniently expressed in terms of cylindrical coordinates - see Section 13.6.

6. No. It depends on the solution of the differential equation (13.6).

7. From the value of the eccentricity e.

8. Kepler observed and recorded the motion of the planets - over a period of 20 years!!

Chapter 14

1. Use Equation (14.0) i.e., $U_{1-2} = \int_{\mathbf{r}_1}^{\mathbf{r}_2} \mathbf{F} \cdot \mathbf{dr} = \int_{s_1}^{s_2} F \cos\theta \, ds$.

2. The principle of work and energy is used to solve kinetic problems that involve *velocity, force and displacement* (since these terms are involved in the equation describing the principle i.e., Equation (14.4)).

3. *Power* is defined as the amount of work performed per unit of time. Once $\mathbf{F}$ and the velocity $\mathbf{v}$ of the point where $\mathbf{F}$ is applied have been found, the power is determined by multiplying the force magnitude by the component of velocity acting in the direction of $\mathbf{F}$ i.e., $P = \mathbf{F} \cdot \mathbf{v} = Fv \cos\theta$.

4. The *mechanical efficiency of a machine* is defined by

$$\epsilon = \frac{\text{power output}}{\text{power input}} \text{ or } \epsilon = \frac{\text{energy output}}{\text{energy input}}$$

Since machines consist of a series of moving parts, frictional forces will always be developed within the machine. As a result, extra energy or power is needed to overcome these forces. Consequently, *the efficiency of a machine is always less than one.*

5. When the work done by a force in moving a particle from one point to another is *independent of the path* followed by the particle, then this force is called a *conservative force.* e.g., weight of a particle or spring force acting on a particle.

6. The work done by the weight of a particle depends only on particle's *vertical displacement*.

7. The *force of friction* exerted *on a moving object* by a fixed surface *depends on the path* of the object i.e., the longer the path, the greater the work. Consequently, frictional forces are *nonconservative*.

8. *Potential energy* is a measure of the amount of work a conservative force will do when it moves from a given position to a reference datum e.g., gravitational potential energy, elastic potential energy (see Section 14.5)

9. Check to see that Equation (14.6) i.e., $\mathbf{F} = -\nabla V$ is satisfied.

10. The conservation of energy equation is used to solve problems involving velocity, displacement and *conservative force systems* - all of which form part of the energy equation (14.7) i.e., $T_1 + V_1 = T_2 + V_2$.

Chapter 15

1. A particle's linear momentum is described by the *vector* $\mathbf{L} = m\mathbf{v}$. It's magnitude is mv and its direction is the same as that of the velocity $\mathbf{v}$.

2. The *principle of linear impulse and momentum* is used to solve problems involving *force, time and velocity*. It provides a *direct means* of obtaining the particle's final velocity $\mathbf{v}_2$ after a specified time period when the particle's initial velocity is known and the forces acting on the particle are either constant or can be expressed *as functions of time*.

3. When the sum of the external impulses acting on a system of particles is zero or negligible (forces causing the impulses are nonimpulsive).

4. Forces causing *negligible* impulses.

5. The solution to the problem can be obtained using the following two equations:

   (a) The conservation of momentum applies to the system of particles:

$$\sum m v_1 = \sum m v_2.$$

(b) The coefficient of restitution e relates the relative velocities of the particles along the line of impact, just before and just after collision.

6. The angular momentum $\mathbf{H}_O$ of a particle about point O is defined as the "moment" of the particle's linear momentum about O. It is sometimes referred to as the moment of momentum.

7. When the angular impulses acting on a particle are all zero during the time t_1 to t_2.

8. False. For example, when the particle is subjected only to a central force. Then the impulsive central force $\mathbf{F}$ is always directed towards point O as the particle moves along the path. Hence, the angular impulse (moment) created by $\mathbf{F}$ about the z-axis passing through point O is always zero and therefore angular momentum of the particle is always conserved about the z-axis but linear momentum is not.

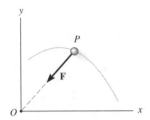

9.

$$m(v_x)_1 + \sum \int_{t_1}^{t_2} F_x dt = m(v_x)_2,$$

$$m(v_y)_1 + \sum \int_{t_1}^{t_2} F_y dt = m(v_y)_2,$$

$$(H_O)_1 + \sum \int_{t_1}^{t_2} M_O\, dt = (H_O)_2.$$

10. This force is internal to the system.

Chapter 16

1. (a) **Translation** Every line segment in the body remains parallel to its original direction during the motion. Specifically, a body can undergo two types of translation:

   i. **Rectilinear Translation.** All points follow parallel straight-line paths.

   ii. **Curvilinear Translation.** All points follow curved paths that are the same shape and are equidistant from one another.

   (b) **Rotation about a Fixed Axis.** All of the particles of the body, except those which lie on the axis of rotation, move along circular paths.

   (c) **General Plane Motion**. The body undergoes a combination of translation *and* rotation.

2. No - angular velocity is a property of the entire body or any *line* in the body.

3. By differentiating the position-coordinate equation $s = f(\theta)$ *with respect to time.*

4. If $s = f(\theta) = \sin^2 \theta$,

$$\frac{ds}{dt} = \frac{d}{dt}\sin^2\theta = \frac{d}{d\theta}\left(\sin^2\theta\right)\cdot\frac{d\theta}{dt}$$
$$= \frac{d}{du}\left(u^2\right)\frac{du}{d\theta}\cdot\frac{d\theta}{dt} \text{ where } u = \sin\theta$$
$$= 2u\left(\cos\theta\right)\dot\theta$$
$$= 2\sin\theta\cos\theta\omega$$
$$= \omega\sin 2\theta$$

5. No - the IC can be at infinity e.g., for a body in pure translation.

6. No because for the IC of zero velocity is not necessarily the IC of zero acceleration. Hence, $\mathbf{a}_{IC} \neq \mathbf{0}$.

7. $\left(\mathbf{v}_{B/A}\right)_{xyz} = 0$ means that point B does not move with respect to point A, for example, if A and B are on the same rigid body.

Chapter 17

1. The *moment of inertia* is a measure of the resistance of a body to *angular acceleration* in the same way that mass is a measure of the body's resistance to *acceleration*.

2. The density $\rho\,(x,\,y,\,z) = \rho = \text{constant}$.

3. Use tables - such as those on the inside back cover of the text.

4. The term $\sum\,(M_k)_P$ represents $I_G\alpha+$ kinetic moments of the components of $m\mathbf{a}_G$ about point P. It arises because we choose to sum moments about a point $P \neq G$ so that the acceleration of the mass center generates "moments" about P. Clearly, these don't exist when moments are summed about G.

5. False. In fact, as per the answer to Q.4 if the moment summation is taken about an arbitrary point $P \neq G$, it is necessary to account for the kinetic moments i.e., we have *either*

$$\sum M_G = 0$$

or summing moments about $P \neq G$:

$$\sum M_P = \sum (M_k)_P$$

6. No - only if the point P coincides with the point through which the axis of rotation passes - for then, by the parallel axis theorem

$$\sum\,(M_k)_P = I_P\alpha$$

If P is any other point, $\sum\,(M_k)_P = I_G\alpha+$ kinetic moments of the components of $m\mathbf{a}_G$ about point P *does not* collapse by the parallel axis theorem to $I_P\alpha$.

7. No - since $\mathbf{a}_{IC} \neq \mathbf{0}$. (!!)

8.

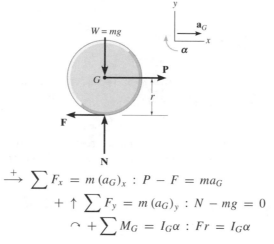

$$\xrightarrow{+}\ \sum F_x = m\,(a_G)_x\ :\ P - F = ma_G$$
$$+\uparrow \sum F_y = m\,(a_G)_y\ :\ N - mg = 0$$
$$\curvearrowleft + \sum M_G = I_G\alpha\ :\ Fr = I_G\alpha$$

If the disk rolls without slipping,

$$a_G = \alpha r$$

We now have 4 linear algebraic equations in the 4 unknowns F, N, α and a_G - this is sufficient information to solve for the 4 unknowns, including a_G, the magnitude of the acceleration of G?

Chapter 18

1. Rigid bodies have mass *and* shape (particles have only mass and are assumed to have no size/shape). Hence, rigid bodies can, in addition, support rotations and hence couple moments. This means that the kinetic energy of a rigid body is made up of *two parts*: kinetic energy of translation *and* kinetic energy of rotation (which involves a moment of inertia). The kinetic energy of rotation does not appear for particles. Similarly, the work done on a rigid body can be composed of work done by external forces *and* work done by external couple moments. Again, the latter does not appear for particles. Otherwise, the symbolic form of the principles of work and energy and conservation of energy are the same for both particles and rigid bodies.

2. $U_{1-2} = \int_s F \cos\theta ds$.

3.

$$U_M = \int_{\theta_1}^{\theta_2} M d\theta$$
$$= M(\theta_2 - \theta_1) \text{ if } M \text{ is constant}$$

4. The principle of work and energy is used to solve kinetic problems that involve *velocity, force and displacement* (since these terms are involved in the equation describing the principle).

5. When the work done by a force in moving a body from one position to another is *independent of the path* followed by the body, then this force is called a *conservative force.* e.g., weight of a body or force developed by an elastic spring.

6. The work done by the weight of a body (considered concentrated at its center of gravity) depends only on body's *vertical displacement.*

7. The *force of friction* exerted *on a moving object* by a fixed surface *depends on the path* of the object i.e., the longer the path, the greater the work. Consequently, frictional forces are *nonconservative.*

8. *Potential energy* is a measure of the amount of work a conservative force will do when it moves from a given position to a reference datum e.g., gravitational potential energy, elastic potential energy (see Section 14.5)

9. Check to see that $\mathbf{F} = -\nabla V$ is satisfied where V is the potential function.

10. The conservation of energy equation is used to solve problems involving velocity, displacement and *conservative force systems* - all of which form part of the conservation of energy equation $T_1 + V_1 = T_2 + V_2$.

Chapter 19

1. The linear momentum of a rigid body is the vector

$$\mathbf{L} = m\mathbf{v}_G$$

with magnitude mv_G (units of $kg \cdot m/s$ or $slug \cdot ft/s$) and a direction defined by $\mathbf{v}_G$.

2. The angular momentum of a rigid body is the vector

$$\mathbf{H}_G = I_G\boldsymbol{\omega}$$

with magnitude $I_G\omega$ (units of $kg \cdot m^2/s$ or $slug \cdot ft^2/s$) and a direction defined by $\boldsymbol{\omega}$, which is always perpendicular to the plane of motion. The angular momentum of the body can also be computed about a point $P \neq G$. In this case:

$$\mathbf{H}_P = I_G\boldsymbol{\omega} + \text{ moment of the linear momentum } m\mathbf{v}_G \text{ about } P$$

3. False. When a body is in translation, the body's linear and angular momentum is given by

$$L = mv_G$$
$$H_A = (d)(mv_G)$$

where A is any point on or off the body. Only in the case where $A \equiv G$ can we say $H_G = 0$

4. False. When a body is in general plane motion, the body's linear and angular momentum is given by

$$L = mv_G$$
$$H_A = I_G\omega + (d)(mv_G)$$

where A is any point on or off the body. Only in the case where $A \equiv G$ can we say $H_G = I_G\omega$.

5. It eliminates the need to include reactive impulses which occur at connections since they are *internal to the system.*

6. A *nonimpulsive force* is one which creates a *small or negligible* impulse.

7. Angular momentum is conserved about the swimmers mass center G since the swimmer's weight has a line of action which passes through G . By tucking his arms and legs in close to his chest, the swimmer decreases his body's moment of inertia and thus increases his angular velocity ($I_G\omega$ must be constant since angular momentum is conserved about G). If he straightens out just before entering the water, his body's moment of inertia is increased and his angular velocity decreases (for smoother entry into the water).

8. No - go back to the example of the diving swimmer in Problem #7. Since the weight of the swimmer's body creates a *linear impulse* during the time of motion, linear momentum *is not conserved* in the downward direction - even though angular momentum *is* conserved about G during the time of motion.

Chapter 20

1. False - see Section 20.1 of text.

2. A "very small" rotation - see Section 20.1 of text.

3. In fact, the angular velocity vector $\boldsymbol{\Omega}$ in Equation (20.0) is the vector that may be used in relating the derivatives in two frames of any arbitrary vector (furthermore, even though, in Equation (20.0), we have used time t as the independent variable, these derivatives may be taken with respect to *any* scalar variable).

4. From the defining Equation (20.0) for $\boldsymbol{\Omega}$, it is clear that angular velocity is a vector relating two *frames*, thus it is meaningless to talk about the angular velocity of a point.

5. Let $\boldsymbol{\Omega}_1$ and $\boldsymbol{\Omega}_2$ be two distinct vectors satisfying Equation (20.0) i.e.,

$$\dot{\mathbf{A}} = \left(\dot{\mathbf{A}}\right)_{xyz} + \boldsymbol{\Omega}_1 \times \mathbf{A}$$
$$\dot{\mathbf{A}} = \left(\dot{\mathbf{A}}\right)_{xyz} + \boldsymbol{\Omega}_2 \times \mathbf{A}$$

for an arbitrary vector $\mathbf{A}$. It then follows that

$$\boldsymbol{\Omega}_1 \times \mathbf{A} - \boldsymbol{\Omega}_2 \times \mathbf{A} = 0$$
$$(\boldsymbol{\Omega}_1 - \boldsymbol{\Omega}_2) \times \mathbf{A} = 0$$

Since $\mathbf{A}$ is arbitrary, we must have that $\boldsymbol{\Omega}_1 = \boldsymbol{\Omega}_2$. If more than two angular velocities are considered, they can be combined into pairs and each pair then coincides by the above result. Hence, the angular velocity vector $\boldsymbol{\Omega}$ satisfying Equation (20.0) is unique.

6. All vectors are three component vectors in three-dimensional kinematics as opposed to two component vectors in plane kinematics.

Chapter 21

1. True

2. Three moments of inertia: I_{xx}, I_{yy} and I_{zz} and six products of inertia (only 3 are independent): $I_{xy} = I_{yx}$, $I_{yz} = I_{zy}$ and $I_{xz} = I_{zx}$

3. The inertia tensor is defined by

$$\begin{pmatrix} I_{xx} & -I_{xy} & -I_{xz} \\ -I_{yx} & I_{yy} & -I_{yz} \\ -I_{zx} & -I_{zy} & I_{zz} \end{pmatrix}$$

It completely characterizes the inertial properties of as body in three-dimensional motion.

4. In the $\omega-$components.

5. No - this is evident from the set of Euler equations (21.0) which is the analogue of the equation $\sum M_G = I_G\omega$ for three-dimensional motion.

6. The interpretation and calculation of kinetic energy - otherwise the symbolism of the principle is the same in the plane and in three-dimensions.

Chapter 22

1. Free vibration occurs when the motion is maintained by gravitational or elastic restoring forces.

2. The differential equation describing undamped free vibrations is $\ddot{x} + p^2 x = 0$. Its general solution and other vibrational characteristics are

$$x(t) = A \sin pt + B \cos pt$$
$$x(t) = C \sin(pt + \phi)$$
$$C = \sqrt{A^2 + B^2} \text{ is the } \textit{amplitude}$$
$$\tau = \frac{2\pi}{p} \text{ is the } \textit{period}$$
$$f = \frac{1}{\tau} \text{ is the } \textit{(natural) frequency}$$

3. When a body or system of connected bodies is given an initial displacement from its equilibrium position and released, it will vibrate with a definite frequency known as the *natural frequency*.

4. Simple harmonic motion describes a situation where acceleration is proportional to displacement:

5. Since in undamped free vibrations the motion is due to *only* gravitational and elastic restoring forces, which are conservative.

6. If, in *forced undamped vibrations* of a system, force or displacement is applied with a frequency close to the natural frequency of the system i.e., $\frac{\omega}{p} \approx 1$, the amplitude of vibration becomes extremely large. This condition is known as *resonance*. This occurs because the force is applied so that it always follows the motion of the vibrating body.

7. There is only one case in which vibrations occur at all. This is the *underdamped system,* in which case λ_1 and λ_2 are both complex and the general solution of (22.2) is given by

$$x(t) = D \left[e^{-\left(\frac{c}{2m}\right)t} \sin(p_d t + \phi) \right]$$

It is clear that this solution, and hence vibrations decay with time.

8. The differential equation which describes the motion of a block and spring system which includes the effects of forced motion *and* induced damping is:

$$m\ddot{x} + c\dot{x} + kx = F_0 \sin \omega t$$

The general solution is of the form

$$x(t) = x_c(t) + x_p(t)$$

where $x_c(t)$ is the general solution of Equation (22.2) (which depends on the values of λ_1 and λ_2 in (22.3) and (22.4) - see (22.5)–(22.7)) and

$$x_p(t) = C' \sin(\omega t - \phi')$$

where the constants C' and ϕ' are given by

$$C' = \frac{\frac{F_0}{k}}{\sqrt{\left[1 - \left(\frac{\omega}{p}\right)^2 \right]^2 + \left[2\left(\frac{c}{c_c}\right)\left(\frac{\omega}{p}\right) \right]^2}},$$

$$\phi' = \tan^{-1} \left[\frac{2\left(\frac{c}{c_c}\right)\left(\frac{\omega}{p}\right)}{1 - \left(\frac{\omega}{p}\right)^2} \right]$$

Hence the solution is in two parts: one part deals with free damped vibrations whereas the other (x_p) deals with the forced component of motion.

PART II

Free-Body Diagram Workbook

1

Basic Concepts in Dynamics

Engineering mechanics is divided into two areas: statics and dynamics. *Statics* deals with the equilibrium of bodies, that is, those that are either at rest (if originally at rest) or move with constant velocity (if originally in motion). *Dynamics* is concerned with the accelerated motion of bodies. The study of dynamics is itself divided into two parts: *kinematics*, which treats only the geometric aspects of motion and *kinetics* which is concerned with the analysis of forces causing the motion. Free-body diagrams play a significant role in solving problems in kinetics.

In mechanics, real bodies (e.g., planets, cars, planes, tables, crates, etc) are represented or *modeled* using certain idealizations which simplify application of the relevant theory. In this book we refer to only two such models:

- **Particle**. A *particle* has a mass but a size/shape that can be neglected. For example, the size of an aircraft is insignificant when compared to the size of the earth and therefore the aircraft can be modeled as a particle when studying its three-dimensional motion in space.
- **Rigid Body**. A *rigid body* represents the next level of sophistication after the particle. That is, a rigid body is a collection of particles which has a size/shape but this size/shape cannot change. In other words, when a body is modeled as a rigid body, we assume that any deformations (changes in shape) are relatively small and can be neglected. For example, the actual deformations occurring in most structures and machines are relatively small so that the rigid body assumption is suitable in these cases.

1.1 Equations of Motion

Equation of Motion for a Particle

When a system of forces acts on a particle, the equation of motion may be written in the form

$$\sum \mathbf{F} = m\mathbf{a} \tag{1.1}$$

where $\sum \mathbf{F}$ is the vector sum of all the external forces acting on the particle and m and $\mathbf{a}$ are, respectively, the mass and acceleration of the particle.

Successful application of the equation of motion (1.1) requires a complete specification of all the known and unknown external forces ($\sum \mathbf{F}$) that act on the object. The best way to account for these is to draw the object's *free-body diagram*: a sketch of the object *freed* from its surroundings showing *all* the (external) forces that act on it. In dynamics problems, since the resultant of these external forces produces the vector $m\mathbf{a}$, in addition to the free-body diagram, a *kinetic diagram* is often used to represent graphically the magnitude and direction of the vector $m\mathbf{a}$. In other words, the equation (1.1) can be represented graphically as:

Free-body Diagram = Kinetic Diagram

Of course, whenever the equation of motion (1.1) is applied, it is required that measurements of the acceleration be made from a *Newtonian* or inertial frame of reference. *Such a coordinate system does not rotate and is either fixed or translates in a given direction with a constant velocity (zero acceleration).* This definition ensures that the particle's acceleration measured by observers in two different inertial frames of reference will always be the *same.*

Equation of Motion for a System of Particles

The equation of motion (1.1) can be extended to include a *system of particles* isolated within an enclosed region in space:

$$\sum \mathbf{F} = m\mathbf{a}_G \tag{1.2}$$

This equation states that the sum of external forces ($\sum \mathbf{F}$) acting on the system of particles is equal to the total mass m of the particles multiplied by the acceleration $\mathbf{a}_G$ of its mass center G. Since, in reality, all particles must have a finite size to possess mass, equation (1.2) justifies application of the equation of motion to a *body* that is represented as a single particle.

Equations of Motion for a Rigid Body

Since rigid bodies, by definition, have a definite size/shape, their motion is governed by *both* translational and rotational quantities. The translational equation of motion for (the mass center of) a rigid body is basically equation (1.2). That is,

$$\sum \mathbf{F} = m\mathbf{a}_G \tag{1.2}$$

In this case, the equation (1.2) states that the sum of all the external forces acting on the body is equal to the body's mass multiplied by the acceleration of its mass center G.

The rotational equation of motion for a rigid body is given by

$$\sum \mathbf{M}_G = I_G \boldsymbol{\alpha} \tag{1.3}$$

which states that the sum of the applied couple moments and moments of all the external forces computed about a body's mass center $G(\sum \mathbf{M}_G)$ is equal to the product of the moment of inertia of the body about an axis passing through $G(I_G)$ and the body's angular acceleration $\boldsymbol{\alpha}$.

Alternatively, equation (1.3) can be re-written in more general form as:

$$\sum \mathbf{M}_P = \sum (M_k)_P \tag{1.4}$$

Here, $\sum \mathbf{M}_P$ represents the sum of the applied couple moments and the external moments taken about a general point P and $\sum (M_k)_P$ represents the sum of the kinetic moments about P, in other words, the sum of $I_G \boldsymbol{\alpha}$ and the moments generated by the components of the vectors $m\mathbf{a}_G$ about the point P.

When applying the equations of motion (1.2)–(1.4), one should always draw a *free-body diagram* in order to account for the terms involved in ($\sum \mathbf{F}$), ($\sum \mathbf{M}_G$) or ($\sum \mathbf{M}_P$). The *kinetic diagram* is also useful in that it accounts graphically for the acceleration components $m\mathbf{a}_G$ and the term $I_G \boldsymbol{\alpha}$ and it is especially convenient when used to determine the components of $m\mathbf{a}_G$ and the moment terms in $\sum (M_k)_P$.

2

Free-Body Diagrams: the Basics

2.1 Free-Body Diagram: Particle

The equation of motion (1.1) is used to analyze the motion of *an object* (modeled as a particle) when subjected to an unbalanced force system. The first step in this analysis is to draw the **free-body diagram** of the object to identify the external forces ($\sum \mathbf{F}$) acting on it. The object's free-body diagram is simply a sketch of the object *freed* from its surroundings showing *all* the (external) forces that *act* on it. The diagram focuses your attention on the object of interest and helps you identify *all* the external forces ($\sum \mathbf{F}$) acting. Once the free-body diagram is drawn, it may be helpful to draw the corresponding *kinetic diagram*. This diagram accounts graphically for the effect of the acceleration components ($m\mathbf{a}$) on the object. Taken together, these diagrams provide (in graphical form) all the information that is needed to write down the equation of motion (1.1).

EXAMPLE 2.1

The 50-kg crate shown in Figure 1, rests on a horizontal plane for which the coefficient of friction is $\mu_k = 0.3$. The crate is subjected to a towing force of magnitude 400N and moves to the right without tipping over. Draw the free-body and kinetic diagrams of the crate.

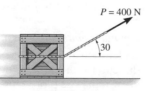

$P = 400$ N

30

Figure 1

Solution

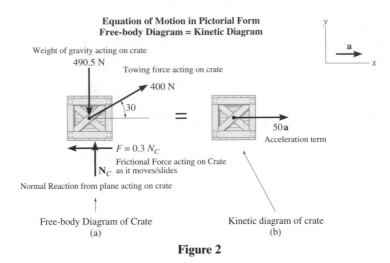

Equation of Motion in Pictorial Form
Free-body Diagram = Kinetic Diagram

Figure 2

The free-body diagram of the crate is shown in Figure 2(a). Notice that once the crate is *separated* or *freed* from the system (= crate + plane), forces which were previously internal to the system become external to the crate. For example, in Figure 2 (a), such a force is the force of friction *acting on the crate*. The kinetic diagram is shown in Figure 2(b). In this case, the diagram shows the effect of the acceleration term $m\mathbf{a}$ on the crate. Taken together, the two diagrams give a pictorial form of the equation of motion (1.1) (or (1.2)). ◄

Next, we present a formal procedure for drawing free-body diagrams for a particle or system of particles.

2.1.1 *Procedure for Drawing a Free-Body Diagram: Particle*

1. *Select* the inertial coordinate system. Most often, rectangular or x, y-coordinates are chosen to analyze problems for which the particle has *rectilinear motion*. If this occurs, one of the axes should extend in the direction of motion.

2. *Identify the object you wish to isolate* from the system. This choice is often dictated by the particular forces of interest.

3. *Draw the outlined shape of the isolated object.* Imagine the object to be isolated or cut free from the system of which it is a part.

4. *Show all external forces acting on the isolated object.* Indicate on this sketch *all* the external forces that act on the object. These forces can be *active forces*, which tend to set the object in motion, or they can be *reactive forces* which are the result of the constraints or supports that prevent motion. This stage is crucial: it may help to trace around the object's boundary, carefully noting each external force acting on it. Don't forget to include the weight of the object (unless it is being intentionally neglected).

5. *Identify and label each external force acting on the (isolated) object.* The forces that are known should be labeled with their known magnitudes and directions. Use letters to represent the magnitudes and arrows to represent the directions of forces that are unknown.

6. *The direction of a force having an unknown magnitude can be assumed.*

7. *The direction and sense* of the particle's acceleration $\mathbf{a}$ should also be established. If the sense of its components is unknown, assume they are in the same direction as the positive inertial coordinate axes. The acceleration may be sketched on the x, y-coordinate system or it may be represented as the $m\mathbf{a}$ vector on the *kinetic diagram*.

2.1.2 Using the Free-Body Diagram: Equations of Motion

The equations of motion (1.1) or (1.2) are used to solve problems which require a relationship between the forces acting on a particle and the accelerated motion they cause. Whenever (1.1) or (1.2) is applied, the unknown force and acceleration components should be identified and an equivalent number of equations should be written. If further equations are required for the solution, kinematics may be considered.

The *free-body diagram* is used to identify the unknown force and the *kinetic diagram* the unknown acceleration components acting on the particle. The subsequent procedure for solving problems once the free-body (and, if necessary, the kinetic) diagram for the particle is established, is therefore as follows:

---

1. If the forces can be resolved directly from the free-body diagram, apply the equations of motion in their scalar component form. For example:

$$\sum F_x = ma_x \quad \text{and} \quad \sum F_y = ma_y \tag{2.1}$$

2. Components are positive if they are directed along a positive axis and negative if they are directed along a negative axis.

3. If the particle contacts a rough surface, it may be necessary to use the frictional equation, which relates the coefficient of kinetic friction to the magnitudes of the frictional and normal forces acting at the surfaces of contact. Remember that the frictional force always acts on the free-body diagram such that it opposes the motion of the particle *relative to the surface it contacts*.

4. If the solution yields a negative result, this indicates the sense of the force is the reverse of that shown/assumed on the free-body diagram.

EXAMPLE 2.2

In Example 2.1, the diagrams established in Figure 2 give us a 'pictorial representation' of all the information we need to apply the equations of motion (2.1) to find the unknown force $\mathbf{N}_C$ and the acceleration $\mathbf{a}$. In fact, taking the positive x-direction to be horizontal ($\rightarrow +$) and the positive y-direction to be vertical ($\uparrow +$), the equations of motion (2.1) when applied to the crate (regarded as a particle — since its shape is not important in the motion under consideration) are:

For the Crate: $\quad \longrightarrow + \sum F_x = ma_x: \quad 400\cos 30° - F = 50a_x$
$$\uparrow + \sum F_y = ma_y: \quad N_C - 490.5 + 400\sin 30° = 0$$

Two equations, 3 unknowns: use the frictional equation to relate F to N_C and obtain a third equation:

$$\text{Frictional Equation (block is sliding):} \quad F = 0.3N_C$$

Solving these three equations yields

$$N_C = 290.5\text{N}, \quad a_x = 5.19 m/s^2 \qquad \textbf{Ans.}$$

The directions of each of the vectors $\mathbf{N}_C$ and $\mathbf{a}$ is shown in the free-body diagram above (Figure 2). ◄

2.2 Free-Body Diagram: Rigid Body

The equations of motion (1.2) and (1.3) (or (1.4)) are used to determine unknown forces, moments and acceleration components acting on an object (modeled as a rigid body) subjected to an unbalanced system of forces and moments. The first step in doing this is again to draw the *free-body diagram* of the object to identify *all of* the external forces and moments acting on it. The procedure for drawing a free-body diagram in this case is much the same as that for a particle with the main difference being that now, because the object has 'size/shape,' it can support also external couple moments and moments of external forces.

2.2.1 Procedure for Drawing a Free-Body Diagram: Rigid Body

1. *Select* the inertial x, y or n, t-coordinate system. This will depend on whether the body is in rectilinear or curvilinear motion.
2. Imagine the body to be isolated or 'cut free' from its constraints and connections and sketch its outlined shape.
3. Identify all the external forces and couple moments that act on the body. Those generally encountered are:
   (a) Applied loadings
   (b) Reactions occurring at the supports or at points of contact with other bodies.
   (c) The weight of the body (applied at the body's center of gravity G)
   (d) Frictional forces
4. The forces and couple moments that are known should be labeled with their proper magnitudes and directions. Letters are used to represent the magnitudes and direction angles of forces and couple moments that are *unknown*. Indicate the dimensions of the body necessary for computing the moments of external forces. In particular, if a force or couple moment has a known line of action but unknown magnitude, the arrowhead which defines the sense of the vector can be assumed. The correctness of the assumed sense will become apparent after solving the equations of motion for the unknown magnitude. By definition, the magnitude of a vector is *always positive*, so that if the solution yields a *negative* scalar, the minus *sign* indicates that the vector's sense is *opposite* to that which was originally assumed.
5. *The direction and sense* of the acceleration of the body's mass center $\mathbf{a}_G$ should also be established. If the sense of its components is unknown, assume they are in the same direction as the positive inertial coordinate axes. The acceleration may be sketched on the x, y-coordinate system or it may be represented as the $m\mathbf{a}_G$ vector on the *kinetic diagram*. This will also be helpful for 'visualizing' the terms needed in the moment sum $\sum (M_k)_P$ since the kinetic diagram accounts graphically for the components $m(a_G)_x$, $m(a_G)_x$ or $m(a_G)_t$, $m(a_G)_n$.

2.2.2 Important Points

- Internal forces are never shown on the free-body diagram since they occur in equal but opposite collinear pairs and therefore cancel each other out.
- The weight of a body is an external force and its effect is shown as a single resultant force acting through the body's center of gravity G.
- *Couple moments* can be placed anywhere on the free-body diagram since they are *free vectors*. Forces can act at any point along their lines of action since they are *sliding vectors*.

EXAMPLE 2.3

Draw the free-body and kinetic diagrams for the 50-kg crate. A force **P** of magnitude $600N$ is applied to the crate as shown. Take the coefficient of kinetic friction to be $\mu_k = 0.2$.

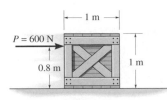

Figure 3

Solution

Here, since the force **P** can cause the crate to either slide or to tip over, we model the crate as a rigid body. This model allows us to account for the effects of moments arising from **P** and any other external forces. We begin by assuming that the crate slides so that the frictional equation yields $F = \mu_k N_C = 0.2N_C$. Also, the normal force N_C acts at O, a distance x (where $0 < x \le 0.5m$) from the crate's center line. Note that the line of action of N_C does not necessarily pass through the mass center $G(x = 0)$, since N_C must counteract the tendency for tipping caused by **P**.

(Note that had we assumed that the crate tips then the normal force N_C would have been assumed to act at the corner point A and the frictional equation would take the form $F \le 0.2N_C$).   ◀

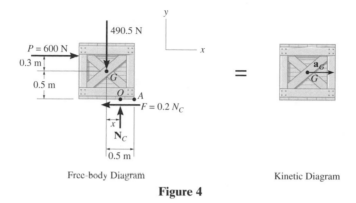

Free-body Diagram   Kinetic Diagram

Figure 4

2.2.3 Using the Free-Body Diagram: Equations of Motion

The procedure for solving kinetic problems for a rigid body once the free-body diagram is established, is as follows:

- Apply the three equations of motion (1.2)–(1.3). To simplify the analysis, the moment equation (1.3) may be replaced by the more general equation (1.4) where the point P is usually located at the intersection of the lines of action of as many unknown forces as possible.
- If the body contacts a rough surface, it may be necessary to use the frictional equation, which relates the coefficient of kinetic friction to the magnitudes of the frictional and normal forces acting at the surfaces of contact. Remember that the frictional force always acts on the free-body diagram such that it *opposes the motion of the body relative to the surface it contacts*.
- Use kinematics if the velocity and position of the body are to be determined.

EXAMPLE 2.4

Find the acceleration of the crate in Example 2.3.

Solution

Using the free-body diagram in Figure 4, the equations of motion are:

$$\longrightarrow + \sum F_x = m(a_G)_x: \quad 600N - 0.2N_C = (50kg)(a_G)_x$$

$$\uparrow + \sum F_y = m(a_G)_y: \quad N_C - 490.5N = 0$$

$$+ \circlearrowright \sum M_G = I_G\alpha: \quad -600N(0.3m) + N_C(x) - 0.2N_C(0.5m) = 0$$

Solving, we obtain

$$\mathbf{N}_C = 490N \uparrow, \quad x = 0.467m, \quad \mathbf{a}_G = 10m/s^2 \longrightarrow \qquad \textbf{Ans.}$$

Since $x = 0.467m < 0.5m$, indeed the crate slides as originally assumed (otherwise the problem would have to be reworked with the assumption that tipping occurred). ◀

3

Problems

3.1 Free-Body Diagrams in Particle Kinetics

Problem 3.1

The sled with load shown has a weight of 50 lb and is acted upon by a force having a variable magnitude $P = 20t$, where P is in pounds and t is in seconds. The coefficient of kinetic friction between the sled and the plane is $\mu_k = 0.3$. Draw the free-body and kinetic diagrams for the sled.

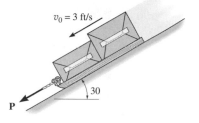

Solution

1. The size/shape of the sled does not affect the (rectilinear) motion under consideration. Consequently, we assume that the sled has *negligible size* so that it can be modelled as a particle.

2. Imagine the sled to be separated or detached from the system (sled + plane).

3. The (detached) sled is subjected to four *external* forces. They are caused by:

   i.   ii.

   iii.   iv.

4. Draw the free-body diagram of the (detached) sled showing all these forces labeled with their magnitudes and directions. Include any other information e.g. angles, lengths etc. which may help when formulating the equations of motion.

5. The acceleration of the sled is down the slope. Show this on a kinetic diagram or on the inertial coordinate system chosen in the free-body diagram.

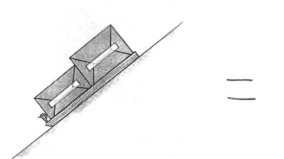 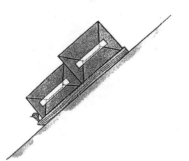

Problem 3.1

The sled with load shown has a weight of 50 lb and is acted upon by a force having a variable magnitude $P = 20t$, where P is in pounds and t is in seconds. The coefficient of kinetic friction between the sled and the plane is $\mu_k = 0.3$. Draw the free-body and kinetic diagrams for the sled.

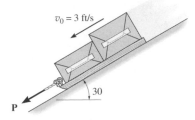

Solution

1. The size/shape of the sled does not affect the (rectilinear) motion under consideration. Consequently, we assume that the sled has *negligible size* so that it can be modelled as a particle.

2. Imagine the sled to be separated or detached from the system (sled + plane).

3. The (detached) sled is subjected to four *external* forces. They are caused by:

      i. Force P   **ii. Sled's weight**

      iii. Frictional force at the surface (sliding)   **iv. Force of surface acting on sled**

4. Draw the free-body diagram of the (detached) sled showing all these forces labeled with their magnitudes and directions. Include any other information e.g. angles, lengths etc. which may help when formulating the equations of motion.

5. The acceleration of the sled is down the slope. Show this on a kinetic diagram or on the inertial coordinate system chosen in the free-body diagram.

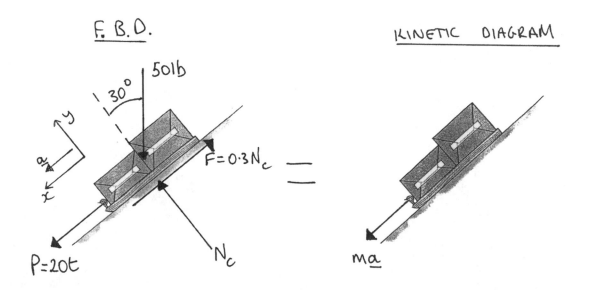

Problem 3.2

The man weighs 180 lb and supports the barbells which have a weight of 100 lb. He lifts the barbells 2 ft in the air in 1.5 secs. Draw the free-body and kinetic diagrams for the man holding the barbells.

Solution

1. The size/shape of the man with barbells does not affect the (rectilinear) motion under consideration. Consequently, we assume that the man with barbells has *negligible size* so that together they can be modelled as a particle.

2. Imagine the man with barbells to be separated or detached from the system (man with barbells + ground).

3. The (detached) man with barbells is subjected to three *external* forces. They are caused by:

   i.   **ii.**

   iii.

4. Draw the free-body diagram of the (detached) man with barbells showing all these forces labeled with their magnitudes and directions. Include any other information e.g. angles, lengths etc. which may help when formulating the equations of motion.

5. The acceleration of the barbells is upward. Show this on a kinetic diagram or on the inertial coordinate system chosen in the free-body diagram.

Problem 3.2

The man weighs 180 lb and supports the barbells which have a weight of 100 lb. He lifts the barbells 2 ft in the air in 1.5 secs. Draw the free-body and kinetic diagrams for the man holding the barbells.

Solution

1. The size/shape of the man with barbells does not affect the (rectilinear) motion under consideration. Consequently, we assume that the man with barbells has *negligible size* so that together they can be modelled as a particle.

2. Imagine the man with barbells to be separated or detached from the system (man with barbells + ground).

3. The (detached) man with barbells is subjected to three *external* forces. They are caused by:

   i. Man's weight   **ii. Weight of barbells**

   iii. (Single) reaction of floor to man (recall man is modelled as a *particle*)

4. Draw the free-body diagram of the (detached) man with barbells showing all these forces labeled with their magnitudes and directions. Include any other information e.g. angles, lengths etc. which may help when formulating the equations of motion.

5. The acceleration of the barbells is upward. Show this on a kinetic diagram or on the inertial coordinate system chosen in the free-body diagram.

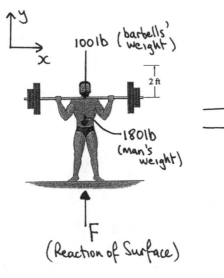

Problem 3.3

The water-park ride consists of an 800-lb sled which slides from rest down the incline and then into the pool. If the frictional resistance on the incline is $F_r = 30$ lb and, in the pool for a short distance $F_r = 80$ lb, draw the free-body and kinetic diagrams for the sled (a) on the incline (b) in the pool. Use these diagrams to determine how fast the sled is travelling when $s = 5$ ft.

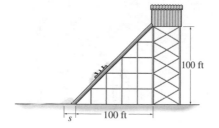

Solution

1. The size/shape of the sled does not affect the (rectilinear) motion under consideration. Consequently, we assume that the sled has *negligible size* so that it can be modelled as a particle.

2. Imagine the sled to be separated or detached from the system (sled + inclined plane or sled + pool).

3. In each case, the (detached) sled is subjected to three *external* forces. They are caused by:

   i.   **ii.**

   iii.

4. Draw the free-body and kinetic diagrams of the (detached) sled showing all these forces labeled with their magnitudes and directions for the sled (a) on incline (b) in pool. Include any other information e.g. angles, lengths etc. which may help when formulating the equations of motion.

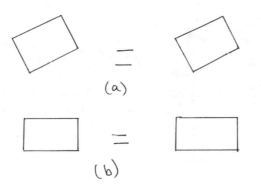

5. In each case, using the coordinate system chosen for the free-body diagrams, write down the equations of motion in the direction of motion and solve for the magnitude of the acceleration:

   Slope: $+ \swarrow \sum F_x = ma_x$:

   Pool: $\longleftarrow + \sum F_x = ma_x$:

6. Use kinematics to find the speed of the sled at $s = 5$ ft.

Problem 3.3

The water-park ride consists of an 800-lb sled which slides from rest down the incline and then into the pool. If the frictional resistance on the incline is $F_r = 30$ lb and, in the pool for a short distance $F_r = 80$ lb, draw the free-body and kinetic diagrams for the sled (a) on the incline (b) in the pool. Use these diagrams to determine how fast the sled is travelling when $s = 5$ ft.

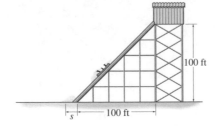

Solution

1. The size/shape of the sled does not affect the (rectilinear) motion under consideration. Consequently, we assume that the sled has *negligible size* so that it can be modelled as a particle.

2. Imagine the sled to be separated or detached from the system (sled + inclined plane or sled + pool).

3. In each case, the (detached) sled is subjected to three *external* forces. They are caused by:

   i. Action of surface on sled   **ii. Sled's weight**

   iii. Frictional force at the surface (sliding)

4. Draw the free-body and kinetic diagrams of the (detached) sled showing all these forces labeled with their magnitudes and directions for the sled (a) on incline (b) in pool. Include any other information e.g. angles, lengths etc. which may help when formulating the equations of motion.

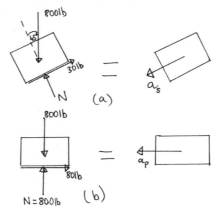

5. In each case, using the coordinate system chosen for the free-body diagrams, write down the equations of motion in the direction of motion and solve for the magnitude of the acceleration:

   Slope: $+ \swarrow \sum F_x = ma_x$: $800 \sin 45° - 30 = \dfrac{800 a_s}{32.2} \implies a_s = 21.561$ ft/s^2

   Pool: $\longleftarrow + \sum F_x = ma_x$: $-80 = \dfrac{800 a_p}{32.2} \implies a_p = -3.22$ ft/s^2

6. Use kinematics to find the speed of the sled at $s = 5$ ft:
   Upon entering the water, the sled has speed v_1 such that $v_1^2 = v_0^2 + 2a_s(s - s_0) = 0 + 2(21.561)\sqrt{20000} = 78.093$ ft/s.
   At $s = 5$ ft, (i.e. 5 ft into the pool)
   $$v_2^2 = v_1^2 + 2a_p(s_2 - s_1) = (78.093)^2 + 2(-3.22)(5 - 0) = 6068.4 \text{ ft/s}$$
   $$v_2 = 77.9 \text{ ft/s} \qquad\qquad\qquad\qquad\qquad\qquad\qquad\qquad\qquad \textbf{Ans.}$$

Problem 3.4

Each of the two blocks has a mass m. The coefficient of kinetic friction at all surfaces of contact is μ. A horizontal force **P** is applied to the bottom block. Draw free-body diagrams for each of the top and bottom blocks.

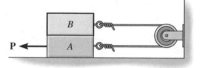

Solution

1. The size/shape of the blocks does not affect the motion under consideration. Consequently, we assume that the blocks have *negligible size* so that they can be modelled as particles.

2. Imagine each block to be separated or detached from the system (two blocks + plane).

3. The (detached) upper block is subjected to four *external* forces. They are caused by:

      i.   ii.

     iii.   iv.

   The (detached) lower block is subjected to six external forces. They are caused by:

      i.   ii.

     iii.   iv.

      v.   vi.

4. Draw the free-body diagrams of each (detached) block showing all these forces labeled with their magnitudes and directions. Include any other information e.g. angles, lengths etc. which may help when formulating the equations of motion.

5. What is the direction of the acceleration of each block? Show this on a kinetic diagram or on the inertial coordinate system chosen in the free-body diagram.

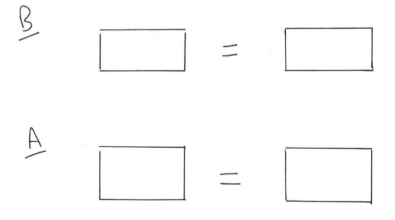

Problem 3.4

Each of the two blocks has a mass m. The coefficient of kinetic friction at all surfaces of contact is μ. A horizontal force **P** is applied to the bottom block. Draw free-body diagrams for each of the top and bottom blocks.

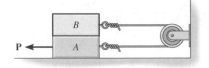

Solution

1. The size/shape of the blocks does not affect the motion under consideration. Consequently, we assume that the blocks have *negligible size* so that they can be modelled as particles.

2. Imagine each block to be separated or detached from the system (two blocks + plane).

3. The (detached) upper block is subjected to four *external* forces. They are caused by:

   i. It's weight

   ii. Cable Tension T

   iii. Friction between blocks

   iv. Reaction from lower block

   The (detached) lower block is subjected to six external forces. They are caused by:

   i. It's weight

   ii. Force P

   iii. Friction at supporting surface

   iv. Friction with upper block

   v. Reaction from surface

   vi. Cable Tension T

4. Draw the free-body diagrams of each (detached) block showing all these forces labeled with their magnitudes and directions. Include any other information e.g. angles, lengths etc. which may help when formulating the equations of motion.

5. What is the direction of the acceleration of each block? Show this on a kinetic diagram or on the inertial coordinate system chosen in the free-body diagram.

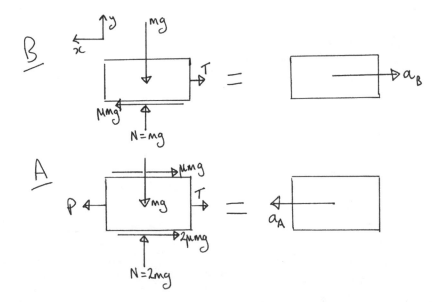

Problem 3.5

The driver attempts to tow the crate which has a weight of 500 lb and which is originally at rest. The coefficient of static friction between the crate and the ground is $\mu_s = 0.4$ and the coefficient of kinetic friction is $\mu_k = 0.3$. Draw free-body and kinetic diagrams for the crate just before and just after it begins to slide.

Solution

1. The size/shape of the crate does not affect the (rectilinear) motion under consideration. Consequently, we assume that the crate has *negligible size* so that it can be modelled as a particle.

2. Imagine the crate to be separated or detached from the system (crate + truck + ground).

3. In each case, the (detached) crate is subjected to four *external* forces. They are caused by:

   i.   ii.

   iii.   iv.

4. Draw the free-body diagram of the (detached) crate in each case, showing all these forces labeled with their magnitudes and directions. Include any other information e.g. angles, lengths etc. which may help when formulating the equations of motion.

5. Draw the corresponding kinetic diagram.

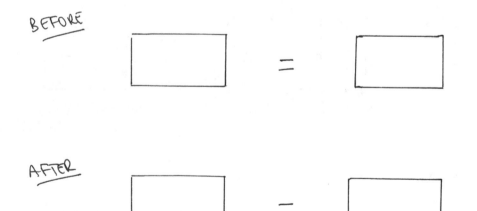

Problem 3.5

The driver attempts to tow the crate which has a weight of 500 lb and which is originally at rest. The coefficient of static friction between the crate and the ground is $\mu_s = 0.4$ and the coefficient of kinetic friction is $\mu_k = 0.3$. Draw free-body and kinetic diagrams for the crate just before and just after it begins to slide.

Solution

1. The size/shape of the crate does not affect the (rectilinear) motion under consideration. Consequently, we assume that the crate has *negligible size* so that it can be modelled as a particle.

2. Imagine the crate to be separated or detached from the system (crate + truck + ground).

3. In each case, the (detached) crate is subjected to four *external* forces. They are caused by:

   i. It's weight   **ii. Tension in rope**

   iii. Friction   **iv. Reaction from surface**

4. Draw the free-body diagram of the (detached) crate in each case, showing all these forces labeled with their magnitudes and directions. Include any other information e.g. angles, lengths etc. which may help when formulating the equations of motion.

5. Draw the corresponding kinetic diagram.

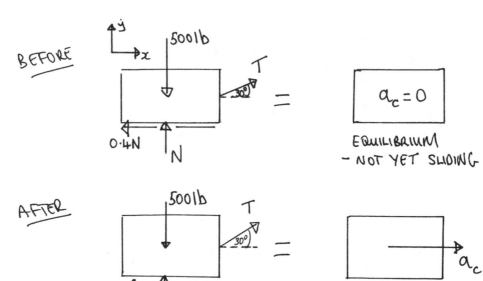

Problem 3.6

The block at B has a mass of 10 kg. Assume the surface at B is smooth. Neglect the mass of the pulleys and cords. Draw free-body and kinetic diagrams for the block at B and use them to formulate an equation of motion which gives a relationship between the acceleration of the block and the tension in the rope attached to B.

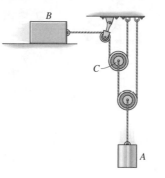

Solution

1. The size/shape of the block does not affect the (rectilinear) motion under consideration. Consequently, we assume that the block has *negligible size* so that it can be modelled as a particle.
2. Imagine the block to be separated or detached from the system.
3. The (detached) block is subjected to three *external* forces. They are caused by:

    i.   **ii.**

    iii.

4. Draw the free-body diagram of the (detached) block showing all these forces labeled with their magnitudes and directions. Include any other information e.g. angles, lengths etc. which may help when formulating the equations of motion.
5. Draw the corresponding kinetic diagram.

6. Using the $xy - axes$ system on the free-body diagram, write down the equation of motion in the x-direction:

$$\underset{\rightarrow}{+} \sum F_x = ma_x:$$

7. Solve for the acceleration of the block:

Problem 3.6

The block at B has a mass of 10 kg. Assume the surface at B is smooth. Neglect the mass of the pulleys and cords. Draw free-body and kinetic diagrams for the block at B and use them to formulate an equation of motion which gives a relationship between the acceleration of the block and the tension in the rope attached to B.

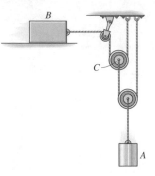

Solution

1. The size/shape of the block does not affect the (rectilinear) motion under consideration. Consequently, we assume that the block has *negligible size* so that it can be modelled as a particle.
2. Imagine the block to be separated or detached from the system.
3. The (detached) block is subjected to three *external* forces. They are caused by:

    i. It's weight   **ii. Tension in rope**

    iii. Reaction from surface

4. Draw the free-body diagram of the (detached) block showing all these forces labeled with their magnitudes and directions. Include any other information e.g. angles, lengths etc. which may help when formulating the equations of motion.
5. Draw the corresponding kinetic diagram.

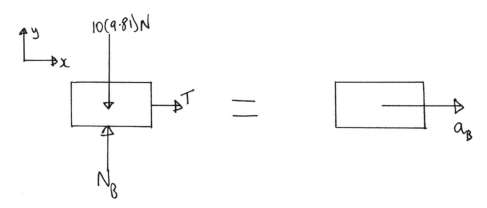

6. Using the xy-axes system on the free-body diagram, write down the equation of motion in the x-direction:

$$\overset{+}{\underset{\rightarrow}{}} \sum F_x = ma_x: \quad T = 10a_B$$

7. Solve for the acceleration of the block:

$$a_B = \frac{T}{10} \text{ m/s}^2 \rightarrow \qquad\qquad\qquad \textbf{Ans.}$$

Problem 3.7

The 400-kg mine car is hoisted up the incline using the cable and motor M. For a short time, the force in the cable has magnitude $F = (3200t^2)N$ where t is in seconds. Draw a free-body diagram for the car and use it to determine the acceleration of the car after 2 seconds.

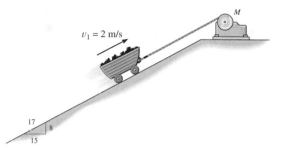

Solution

1. The size/shape of the car does not affect the (rectilinear) motion under consideration. Consequently, we assume that the car has *negligible size* so that it can be modelled as a particle.
2. Imagine the car to be separated or detached from the system.
3. The (detached) car is subjected to three *external* forces. They are caused by:

   i.   **ii.**

   iii.

4. Draw the free-body diagram of the (detached) car showing all these forces labeled with their magnitudes and directions. Include any other information e.g. angles, lengths etc. which may help when formulating the equations of motion.
5. Draw the corresponding kinetic diagram.

6. Using the $xy - axes$ system on the free-body diagram, write down the equation of motion in the x-direction:

   $$\nearrow + \sum F_x = ma_x:$$

7. Solve for the acceleration of the car:

Problem 3.7

The 400-kg mine car is hoisted up the incline using the cable and motor M. For a short time, the force in the cable has magnitude $F = (3200t^2)N$ where t is in seconds. Draw a free-body diagram for the car and use it to determine the acceleration of the car after 2 seconds.

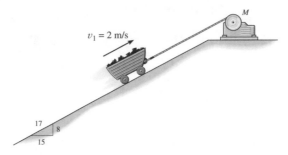

Solution

1. The size/shape of the car does not affect the (rectilinear) motion under consideration. Consequently, we assume that the car has *negligible size* so that it can be modelled as a particle.

2. Imagine the car to be separated or detached from the system.

3. The (detached) car is subjected to three *external* forces. They are caused by:

   i. It's weight   **ii. Force in the cable**

   iii. Reaction from surface

4. Draw the free-body diagram of the (detached) car showing all these forces labeled with their magnitudes and directions. Include any other information e.g. angles, lengths etc. which may help when formulating the equations of motion.

5. Draw the corresponding kinetic diagram.

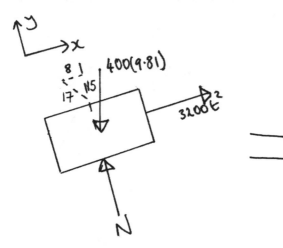
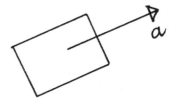

6. Using the $xy-axes$ system on the free-body diagram, write down the equation of motion in the x-direction:

$$\nearrow + \sum F_x = ma_x: \quad 3200t^2 - 400(9.81)\left(\frac{8}{17}\right) = 400a$$

7. Solve for the acceleration of the car:

$$a(t) = (8t^2 - 4.616) \text{ m/s}^2; \quad a(2) = 27.384 \text{ m/s}^2 \rightarrow \qquad \text{**Ans.**}$$

Problem 3.8

Block *A* has a weight of 8 lb and block *B* has a weight of 6 lb. They rest on a surface for which the coefficient of kinetic friction is $\mu_k = 0.2$. If the spring has a stiffness of $k = 20$ lb/ft, and it is compressed 0.2 ft, draw free-body diagrams for both blocks and use them to determine the acceleration of each block just after they are released.

Solution

1. The size/shape of the blocks does not affect the motion under consideration. Consequently, we assume that the blocks have *negligible size* so that they can be modelled as particles.

2. Imagine each block to be separated or detached from the system.

3. Block *A* is subjected to four *external* forces. They are caused by:

   i.   ii.

   iii.   iv.

   Block *B* is subjected to four external forces. They are caused by:

   i.   ii.

   iii.   iv.

4. Draw the free-body diagrams of each (detached) block showing all these forces labeled with their magnitudes and directions. Include any other information e.g. angles, lengths etc. which may help when formulating the equations of motion. What is the direction of the acceleration vector for each block. Show this on a kinetic diagram or on the inertial coordinate system chosen in each free-body diagram.

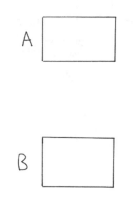

5. Using the xy-axes system on the free-body diagram, write down the equation of motion in the x-direction for each block:

   Block *A*: $\xleftarrow{+} \sum F_x = ma_x$:

   Block *B*: $\xrightarrow{+} \sum F_x = ma_x$:

6. Solve for the acceleration in each case:

Problem 3.8

Block A has a weight of 8 lb and block B has a weight of 6 lb. They rest on a surface for which the coefficient of kinetic friction is $\mu_k = 0.2$. If the spring has a stiffness of $k = 20$ lb/ft, and it is compressed 0.2 ft, draw free-body diagrams for both blocks and use them to determine the acceleration of each block just after they are released.

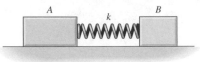

Solution
1. The size/shape of the blocks does not affect the motion under consideration. Consequently, we assume that the blocks have *negligible size* so that they can be modelled as particles.

2. Imagine each block to be separated or detached from the system.

3. Block A is subjected to four *external* forces. They are caused by:

   i. It's weight   **ii. Spring force**

   iii. Friction   **iv. Reaction from surface**

   Block B is subjected to four external forces. They are caused by:

   i. It's weight   **ii. Spring force**

   iii. Friction   **iv. Reaction from surface**

4. Draw the free-body diagrams of each (detached) block showing all these forces labeled with their magnitudes and directions. Include any other information e.g. angles, lengths etc. which may help when formulating the equations of motion. What is the direction of the acceleration vector for each block. Show this on a kinetic diagram or on the inertial coordinate system chosen in each free-body diagram.

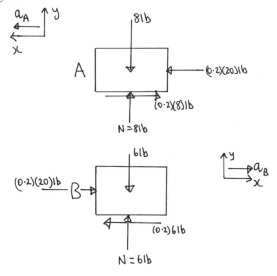

5. Using the xy-axes system on the free-body diagram, write down the equation of motion in the x-direction for each block:

   Block A: $\overset{+}{\leftarrow} \sum F_x = ma_x$: $4 - 1.6 = \dfrac{8}{32.2}a_A$

   Block B: $\overset{+}{\rightarrow} \sum F_x = ma_x$: $4 - 1.2 = \dfrac{6}{32.2}a_B$

6. Solve for the acceleration in each case:

$$a_A = 9.66 \text{ ft/s}^2 \leftarrow, \quad a_B = 15.0 \text{ ft/s}^2 \rightarrow \qquad\qquad \textbf{Ans.}$$

Problem 3.9

When crossing an intersection, a motorcyclist encounters the slight bump or crown caused by the intersecting road. The crest of the bump has a radius of curvature of $\rho = 50$ ft. Draw free-body and kinetic diagrams for the motorcycle with rider. Use these diagrams to formulate equations of motion for the motorcycle with rider and find the maximum constant speed he can travel without leaving the surface of the road. Neglect the size of the motorcycle and the rider in the calculation. The rider and his motorcycle have a total weight of 450 lb.

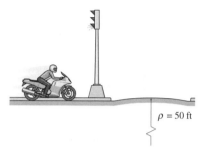

$\rho = 50$ ft

Solution

1. The motorcycle and rider have *negligible size* so that together they can be modelled as a particle.
2. Imagine the motorcycle and rider to be separated or detached from the system.
3. The (detached) motorcycle and rider is subjected to four *external* forces. They are caused by:

   i.   ii.

   iii.   iv.

4. Draw the free-body diagram of the (detached) motorcycle and rider (at the instant he encounters the bump) showing all these forces labeled with their magnitudes and directions. Include any other information e.g. angles, lengths etc. which may help when formulating the equations of motion. Which information given in the question suggests you use a $nt - coordinate$ system as the chosen inertial system? Show the corresponding acceleration components on a kinetic diagram or on the inertial coordinate system chosen in the free-body diagram.

 —

5. Using the nt-axes system on the free-body diagram, write down the equation of motion in the n-direction:

   $$+\downarrow \sum F_n = ma_n:$$

6. Solve for the acceleration component a_n under the appropriate conditions and hence the required speed v of the motorcycle:

Problem 3.9

When crossing an intersection, a motorcyclist encounters the slight bump or crown caused by the intersecting road. The crest of the bump has a radius of curvature of $\rho = 50$ ft. Draw free-body and kinetic diagrams for the motorcycle with rider. Use these diagrams to formulate equations of motion for the motorcycle with rider and find the maximum constant speed he can travel without leaving the surface of the road. Neglect the size of the motorcycle and the rider in the calculation. The rider and his motorcycle have a total weight of 450 lb.

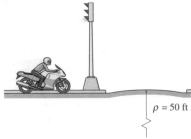

$\rho = 50$ ft

Solution

1. The motorcycle and rider have *negligible size* so that together they can be modelled as a particle.
2. Imagine the motorcycle and rider to be separated or detached from the system.
3. The (detached) motorcycle and rider is subjected to four *external* forces. They are caused by:

    i. Total weight   **ii. Friction at surface**

    iii. Drag   **iv. Reaction from surface**

4. Draw the free-body diagram of the (detached) motorcycle and rider (at the instant he encounters the bump) showing all these forces labeled with their magnitudes and directions. Include any other information e.g. angles, lengths etc. which may help when formulating the equations of motion. Which information given in the question suggests you use a $nt-coordinate$ system as the chosen inertial system? Show the corresponding acceleration components on a kinetic diagram or on the inertial coordinate system chosen in the free-body diagram.

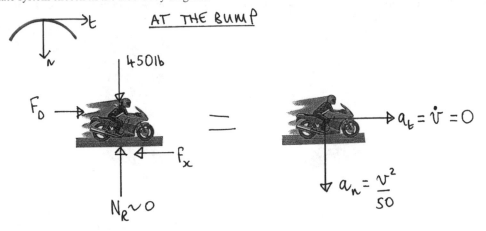

5. Using the nt-axes system on the free-body diagram, write down the equation of motion in the n-direction:

$$+ \downarrow \sum F_n = ma_n: \quad 450 - N_R = \frac{450}{32.2}a_n$$

6. Solve for the acceleration component a_n under the appropriate conditions and hence the required speed v of the motorcycle:

$$\text{Let } N_R = 0 \text{ and } a_n = \frac{v^2}{50} \Rightarrow v = 40.1 \text{ ft/s} \qquad\qquad \textbf{Ans.}$$

Problem 3.10

The 2-kg spool S fits loosely on the inclined rod for which the coefficient of static friction is $\mu_s = 0.2$. If the spool is located 0.25 m from A, use a free-body diagram of the spool to determine the minimum constant speed the spool can have so that it does not slip down the rod.

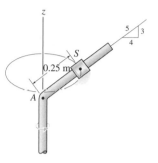

Solution

1. The spool has *negligible size* so that it can be modelled as a particle.
2. Imagine the spool to be separated or detached from the system.
3. The (detached) spool is subjected to three *external* forces. They are caused by:

   i.   ii.

   iii.

4. Draw the free-body diagram of the (detached) spool showing all these forces labeled with their magnitudes and directions. Include any other information e.g. angles, lengths etc. which may help when formulating the equations of motion. Which information given in the question suggests you use a nt-coordinate system as the chosen inertial system? Show the corresponding acceleration components on a kinetic diagram or on the inertial coordinate system chosen in the free-body diagram.

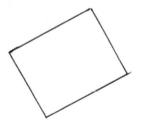

5. Using the nt-axes system on the free-body diagram, write down the equations of motion in the n and t-directions:

   $$\xleftarrow{+} \sum F_n = ma_n:$$

   $$+\uparrow \sum F_t = ma_t:$$

6. Solve for the required speed v of the spool:

Problem 3.10

The 2-kg spool S fits loosely on the inclined rod for which the coefficient of static friction is $\mu_s = 0.2$. If the spool is located 0.25 m from A, use a free-body diagram of the spool to determine the minimum constant speed the spool can have so that it does not slip down the rod.

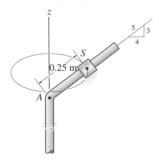

Solution

1. The spool has *negligible size* so that it can be modelled as a particle.
2. Imagine the spool to be separated or detached from the system.
3. The (detached) spool is subjected to three *external* forces. They are caused by:

   i. **It's weight**   ii. **Reaction from surface**

   iii. **Friction**

4. Draw the free-body diagram of the (detached) spool showing all these forces labeled with their magnitudes and directions. Include any other information e.g. angles, lengths etc. which may help when formulating the equations of motion. Which information given in the question suggests you use a $nt - coordinate$ system as the chosen inertial system? Show the corresponding acceleration components on a kinetic diagram or on the inertial coordinate system chosen in the free-body diagram.

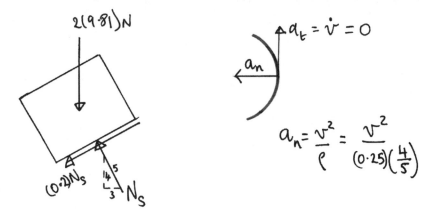

5. Using the nt-axes system on the free-body diagram, write down the equations of motion in the n and t-directions:

$$\overset{+}{\leftarrow} \sum F_n = ma_n: \quad N_s\left(\frac{3}{5}\right) + 0.2N_s\left(\frac{4}{5}\right) = 2a_n$$

$$+ \uparrow \sum F_t = ma_t: \quad N_s\left(\frac{4}{5}\right) - 0.2N_s\left(\frac{3}{5}\right) - 2(9.81) = 2a_t$$

6. Solve for the required speed v of the spool:

$$\text{Set } a_n = \frac{v^2}{0.2}, \quad a_t = 0(\dot{v} = 0) \text{ and obtain: } N_s = 28.85N, \quad v = 1.48 \text{ m/s} \qquad \textbf{Ans.}$$

3.2 Free-Body Diagrams in Rigid Body Kinetics

Problem 3.11

Draw the free-body and kinetic diagrams of the 2-lb bottle, with center of gravity at G, resting on the check-out conveyor. Use these diagrams to write down the equations of motion for the bottle.

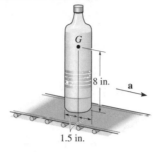

Solution

1. Imagine the bottle to be separated or detached from the system.
2. The bottle is subjected to three *external* forces (don't forget the weight!). They are caused by:

    i.   **ii.**

    iii.

3. Draw the free-body diagram of the (detached) bottle showing all these forces labeled with their magnitudes and directions. The line of action of the force of the belt on the bottle will vary depending on whether the bottle will slip or tip. This should be clear from your free-body diagram. Include any other relevant information e.g. lengths, angles etc. which may help when formulating the equations of motion (including the moment equation) for the bottle.
4. Draw the corresponding kinetic diagram.

5. Using the inertial coordinate system chosen on the free-body diagram and a suitably chosen point O, write down the equations of motion:

$$\rightarrow + \sum F_x = m(a_G)_x:$$

$$+ \uparrow \sum F_y = m(a_G)_y:$$

$$\circlearrowleft \rightarrow + \sum M_O = \sum(M_k)_O:$$

Problem 3.11

Draw the free-body and kinetic diagrams of the 2-lb bottle, with center of gravity at G, resting on the check-out conveyor. Use these diagrams to write down the equations of motion for the bottle.

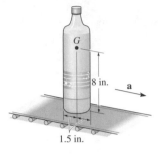

8 in.

a

1.5 in.

Solution

1. Imagine the bottle to be separated or detached from the system.
2. The bottle is subjected to three *external* forces (don't forget the weight!). They are caused by:

   i. It's weight **ii. Friction**

   iii. Reaction from surface

3. Draw the free-body diagram of the (detached) bottle showing all these forces labeled with their magnitudes and directions. The line of action of the force of the belt on the bottle will vary depending on whether the bottle will slip or tip. This should be clear from your free-body diagram. Include any other relevant information e.g. lengths, angles etc. which may help when formulating the equations of motion (including the moment equation) for the bottle.
4. Draw the corresponding kinetic diagram.

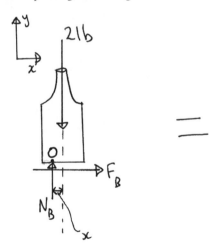

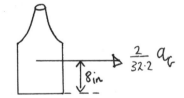

5. Using the inertial coordinate system chosen on the free-body diagram and a suitably chosen point O:

$$\rightarrow + \sum F_x = m(a_G)_x: \quad F_B = \frac{2}{32.2} a_G$$

$$+ \uparrow \sum F_y = m(a_G)_y: \quad N_B 0 - 2 = 0$$

$$\circlearrowleft + \sum M_O = \sum (M_k)_O: \quad 2x = \frac{2}{32.2} a_G (8)$$

Problem 3.12

Draw the free-body and kinetic diagrams of the 200 lb door with center of gravity at G, if a man pushes on it at C with a horizontal force with magnitude F. There are rollers at A and B.

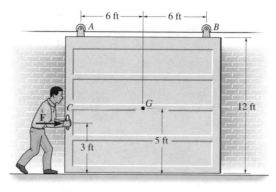

Solution

1. Imagine the door to be separated or detached from the system.
2. The door is subjected to four *external* forces (don't forget the weight!). They are caused by:

   i.   **ii.**

   iii.   **iv.**

3. Draw the free-body diagram of the (detached) door showing all these forces labeled with their magnitudes and directions. Include any other relevant information e.g. lengths, angles etc. which may help when formulating the equations of motion (including the moment equation) for the door.
4. Draw the corresponding kinetic diagram.

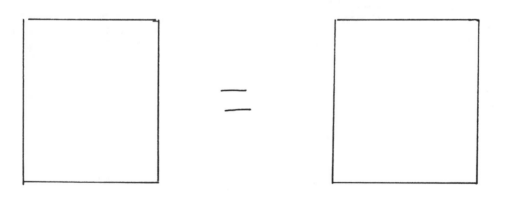

Problem 3.12

Draw the free-body and kinetic diagrams of the 200 lb door with center of gravity at G, if a man pushes on it at C with a horizontal force with magnitude F. There are rollers at A and B.

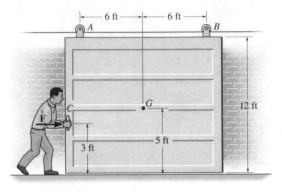

Solution

1. Imagine the door to be separated or detached from the system.

2. The door is subjected to four *external* forces (don't forget the weight!). They are caused by:

   i. It's weight   ii. Roller at A

   iii. Roller at B iv. Force F

3. Draw the free-body diagram of the (detached) door showing all these forces labeled with their magnitudes and directions. Include any other relevant information e.g. lengths, angles etc. which may help when formulating the equations of motion (including the moment equation) for the door.

4. Draw the corresponding kinetic diagram.

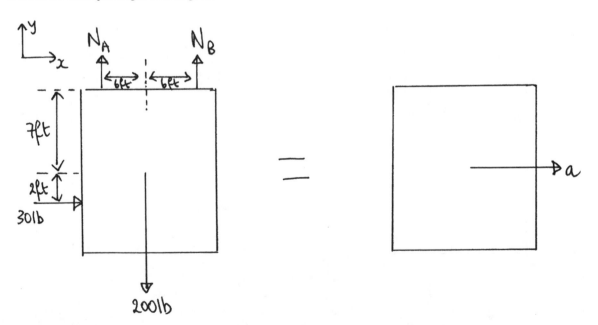

Problem 3.13

The jet has a total mass of 22 Mg and a center of mass at G. Initially at take-off, the engines provide a thrust $2T = 4$ kN and $T' = 1.5$ kN. Draw the free-body and kinetic diagrams of the jet. Neglect the mass of the wheels and, due to low velocity, neglect any lift caused by the wings. There are *two* wing wheels at B and one nose wheel at A.

Solution

1. Imagine the jet to be separated or detached from the system.
2. The jet is subjected to five *external* forces (don't forget the weight!). They are caused by:

      i.   **ii.**

     iii.   **iv.**

      v.

3. Draw the free-body diagram of the (detached) jet showing all these forces labeled with their magnitudes and directions. Include any other relevant information e.g. lengths, angles etc. which may help when formulating the equations of motion (including the moment equation) for the jet.
4. Draw the corresponding kinetic diagram.

Problem 3.13

The jet has a total mass of 22 Mg and a center of mass at G. Initially at take-off, the engines provide a thrust $2T = 4$ kN and $T' = 1.5$ kN. Draw the free-body and kinetic diagrams of the jet. Neglect the mass of the wheels and, due to low velocity, neglect any lift caused by the wings. There are *two* wing wheels at B and one nose wheel at A.

Solution

1. Imagine the jet to be separated or detached from the system.

2. The jet is subjected to five *external* forces (don't forget the weight!). They are caused by:

    i. It's weight   **ii. Reaction at** A

    iii. Reactions at B **iv. Thrust of magnitude** $2T$

    v. Thrust T′

3. Draw the free-body diagram of the (detached) jet showing all these forces labeled with their magnitudes and directions. Include any other relevant information e.g. lengths, angles etc. which may help when formulating the equations of motion (including the moment equation) for the jet.

4. Draw the corresponding kinetic diagram.

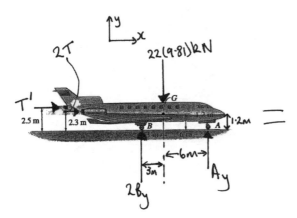

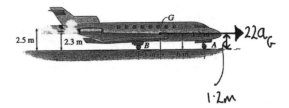

Problem 3.14

The top truck has a mass of 1.75 Mg and a center of mass at G. It is tied to the transport using a chain DE. The transport accelerates at 2 m/s. Draw the free-body and kinetic diagrams of the top truck.

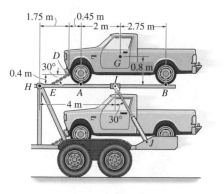

Solution

1. Imagine the top truck to be separated or detached from the system.
2. The truck is subjected to four *external* forces (don't forget the weight!). They are caused by:

   i.   ii.

   iii.   iv.

3. Draw the free-body diagram of the (detached) truck showing all these forces labeled with their magnitudes and directions. Include any other relevant information e.g. lengths, angles etc. which may help when formulating the equations of motion (including the moment equation) for the truck.
4. Draw the corresponding kinetic diagram.

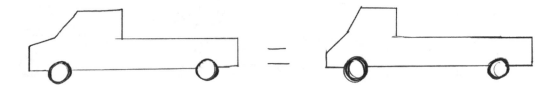

Problem 3.14

The top truck has a mass of 1.75 Mg and a center of mass at G. It is tied to the transport using a chain DE. The transport accelerates at 2 m/s. Draw the free-body and kinetic diagrams of the top truck.

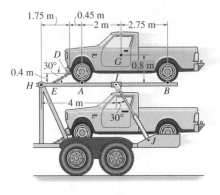

Solution

1. Imagine the top truck to be separated or detached from the system.

2. The truck is subjected to four *external* forces (don't forget the weight!). They are caused by:

   i. It's weight   **ii. Reactions at A**

   iii. Reactions at B   **iv. Chain DE**

3. Draw the free-body diagram of the (detached) truck showing all these forces labeled with their magnitudes and directions. Include any other relevant information e.g. lengths, angles etc. which may help when formulating the equations of motion (including the moment equation) for the truck.

4. Draw the corresponding kinetic diagram.

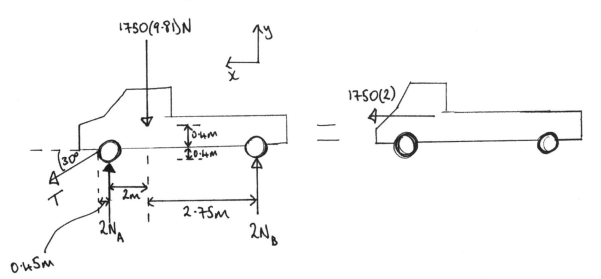

Problem 3.15

The drop gate at the end of the trailer has a mass of 1.25 Mg and mass center at G. It is supported by the cable AB and hinge at C. The truck begins to accelerate at 5 m/s^2. Draw the free-body and kinetic diagrams of the drop gate.

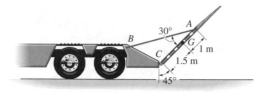

Solution

1. Imagine the drop gate to be separated or detached from the system.
2. The gate is subjected to four *external* forces (don't forget the weight!). They are caused by:

    i.   **ii.**

    iii.   **iv.**

3. Draw the free-body diagram of the (detached) gate showing all these forces labeled with their magnitudes and directions. Include any other relevant information e.g. lengths, angles etc. which may help when formulating the equations of motion (including the moment equation) for the gate.
4. Draw the corresponding kinetic diagram.

Problem 3.15

The drop gate at the end of the trailer has a mass of 1.25 Mg and mass center at G. It is supported by the cable AB and hinge at C. The truck begins to accelerate at 5 m/s^2. Draw the free-body and kinetic diagrams of the drop gate.

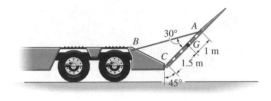

Solution

1. Imagine the drop gate to be separated or detached from the system.
2. The gate is subjected to four *external* forces (don't forget the weight!). They are caused by:

   i. It's weight   **ii. Two reactions at C**

   iii. Cable AB

3. Draw the free-body diagram of the (detached) gate showing all these forces labeled with their magnitudes and directions. Include any other relevant information e.g. lengths, angles etc. which may help when formulating the equations of motion (including the moment equation) for the gate.
4. Draw the corresponding kinetic diagram.

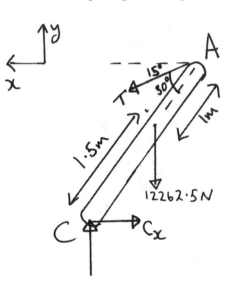

Problem 3.16

The sports car has a weight of 4500 lb and center of gravity at G. It starts fom rest causing the rear wheels to slip as it accelerates. The coefficient of kinetic friction at the road is $\mu_k = 0.3$. Draw the free-body and kinetic diagrams of the car. Neglect the mass of the wheels.

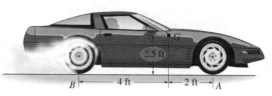

Solution

1. Imagine the car to be separated or detached from the system.
2. The car is subjected to four *external* forces. They are caused by:

   i.   ii.

   iii.   iv.

3. Draw the free-body diagram of the (detached) car showing all these forces labeled with their magnitudes and directions. Include any other relevant information e.g. lengths, angles etc. which may help when formulating the equations of motion (including the moment equation) for the car.
4. Draw the corresponding kinetic diagram.

Problem 3.16

The sports car has a weight of 4500 lb and center of gravity at G. It starts fom rest causing the rear wheels to slip as it accelerates. The coefficient of kinetic friction at the road is $\mu_k = 0.3$. Draw the free-body and kinetic diagrams of the car. Neglect the mass of the wheels.

Solution

1. Imagine the car to be separated or detached from the system.
2. The car is subjected to four *external* forces. They are caused by:

   i. **It's weight**   ii. **Reactions at A**

   iii. **Reactions at B**   iv. **Friction at rear wheels**

3. Draw the free-body diagram of the (detached) car showing all these forces labeled with their magnitudes and directions. Include any other relevant information e.g. lengths, angles etc. which may help when formulating the equations of motion (including the moment equation) for the car.
4. Draw the corresponding kinetic diagram.

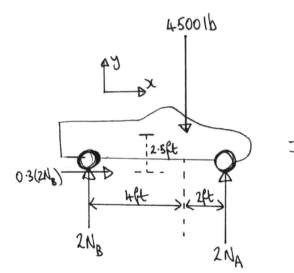

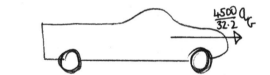

Problem 3.17

The drum truck supports the 600 lb drum that has a center of gravity at G. The operator pushes it forward with a horizontal force of 20 lb. Draw free-body and kinetic diagrams for the drum truck. Neglect the mass of the (4) wheels.

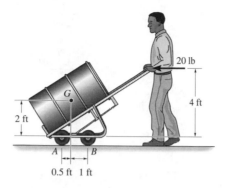

Solution

1. Imagine the drum truck to be separated or detached from the system.
2. The drum truck is subjected to four *external* forces. They are caused by:

   i.   **ii.**

   iii.   **iv.**

3. Draw the free-body diagram of the (detached) truckshowing all these forces labeled with their magnitudes and directions. Include any other relevant information e.g. lengths, angles etc. which may help when formulating the equations of motion (including the moment equation) for the truck.
4. Draw the corresponding kinetic diagram.

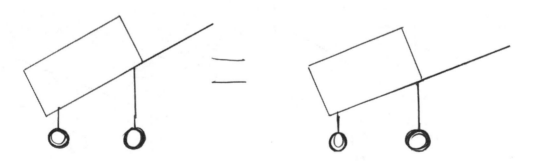

Problem 3.17

The drum truck supports the 600 lb drum that has a center of gravity at G. The operator pushes it forward with a horizontal force of 20 lb. Draw free-body and kinetic diagrams for the drum truck. Neglect the mass of the (4) wheels.

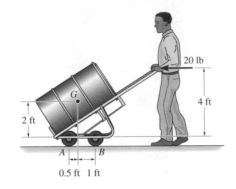

Solution

1. Imagine the drum truck to be separated or detached from the system.
2. The drum truck is subjected to four *external* forces. They are caused by:

      i. **It's weight**
      ii. **Reactions at A**
      iii. **Reactions at B**
      iv. **Force of magnitude 20 lb**

3. Draw the free-body diagram of the (detached) truckshowing all these forces labeled with their magnitudes and directions. Include any other relevant information e.g. lengths, angles etc. which may help when formulating the equations of motion (including the moment equation) for the truck.
4. Draw the corresponding kinetic diagram.

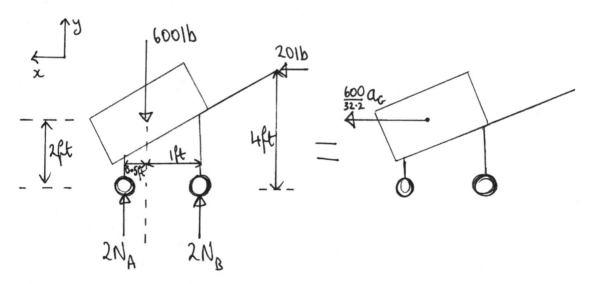

Problem 3.18

The arched pipe has a mass of 80 kg and rests on the surface of the platform. As it is hoisted from one level to the next, $\alpha = 0.25$ rad/s^2 and $\omega = 0.5$ rad/s at the instant $\theta = 30°$. The pipe does not slip. Draw the free-body and kinetic diagrams of the pipe at this instant.

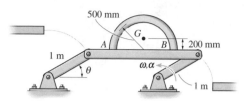

Solution

1. Imagine the pipe to be separated or detached from the system.
2. The pipe is subjected to five *external* forces. They are caused by:

   i.   ii.
   iii.   iv.
   v.

3. Draw the free-body diagram of the (detached) pipe showing all these forces labeled with their magnitudes and directions. Include any other relevant information e.g. lengths, angles etc. which may help when formulating the equations of motion (including the moment equation) for the pipe.
4. Draw the corresponding kinetic diagram.

Problem 3.18

The arched pipe has a mass of 80 kg and rests on the surface of the platform. As it is hoisted from one level to the next, $\alpha = 0.25$ rad/s^2 and $\omega = 0.5$ rad/s at the instant $\theta = 30°$. The pipe does not slip. Draw the free-body and kinetic diagrams of the pipe at this instant.

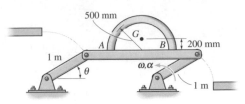

Solution

1. Imagine the pipe to be separated or detached from the system.
2. The pipe is subjected to five *external* forces. They are caused by:

   i. **It's weight**   ii. **Reaction at** A

   iii. **Reaction at** B iv. **Friction at** A

   v. **Friction at** B

3. Draw the free-body diagram of the (detached) pipe showing all these forces labeled with their magnitudes and directions. Include any other relevant information e.g. lengths, angles etc. which may help when formulating the equations of motion (including the moment equation) for the pipe.

4. Draw the corresponding kinetic diagram.

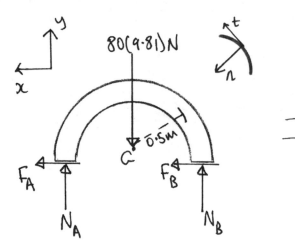

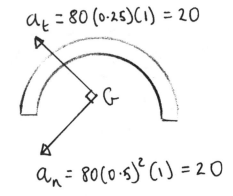

Problem 3.19

The desk has a weight of 75 lb and a center of gravity at G. A man pushes on it at C with a force with magnitude $F = 60$ lb. The coefficient of kinetic friction at A and B is $\mu_k = 0.2$. Draw the free-body and kinetic diagrams of the desk.

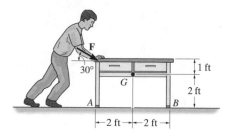

Solution

1. Imagine the desk to be separated or detached from the system.

2. The desk is subjected to six *external* forces. They are caused by:

   i.   ii.

   iii.   iv.

   v.   vi.

3. Draw the free-body diagram of the (detached) desk showing all these forces labeled with their magnitudes and directions. Include any other relevant information e.g. lengths, angles etc. which may help when formulating the equations of motion (including the moment equation) for the desk.

4. Draw the corresponding kinetic diagram.

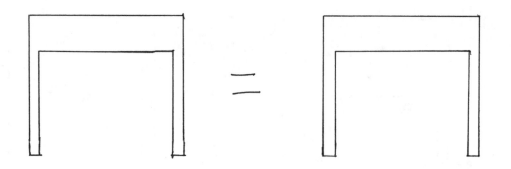

Problem 3.19

The desk has a weight of 75 lb and a center of gravity at G. A man pushes on it at C with a force with magnitude $F = 60$ lb. The coefficient of kinetic friction at A and B is $\mu_k = 0.2$. Draw the free-body and kinetic diagrams of the desk.

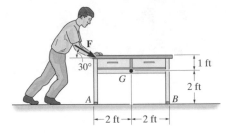

Solution

1. Imagine the desk to be separated or detached from the system.
2. The desk is subjected to six *external* forces. They are caused by:

   i. **It's weight**

   iii. **Reaction at B**

   v. **Friction at B**

   ii. **Reaction at A**

   iv. **Friction at A**

   vi. **Applied force F**

3. Draw the free-body diagram of the (detached) desk showing all these forces labeled with their magnitudes and directions. Include any other relevant information e.g. lengths, angles etc. which may help when formulating the equations of motion (including the moment equation) for the desk.
4. Draw the corresponding kinetic diagram.

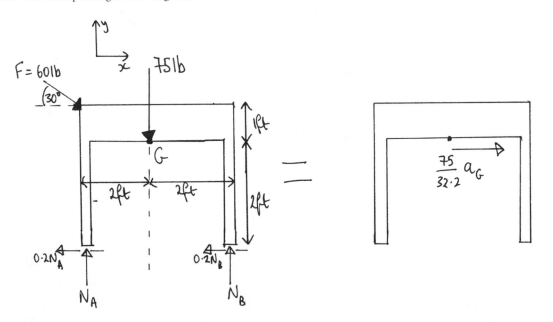

Problem 3.20

The smooth 180-lb pipe has a length of 20 ft and a negligible diameter. It is carried on a truck as shown. Draw the free-body and kinetic diagrams of the pipe.

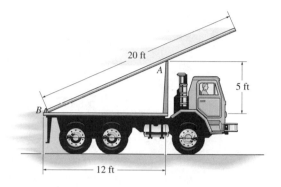

Solution

1. Imagine the pipe to be separated or detached from the system.
2. The pipe is subjected to four *external* forces. They are caused by:

   i.   ii.

   iii.   iv.

3. Draw the free-body diagram of the (detached) pipe showing all these forces labeled with their magnitudes and directions. Include any other relevant information e.g. lengths, angles etc. which may help when formulating the equations of motion (including the moment equation) for the pipe.
4. Draw the corresponding kinetic diagram.

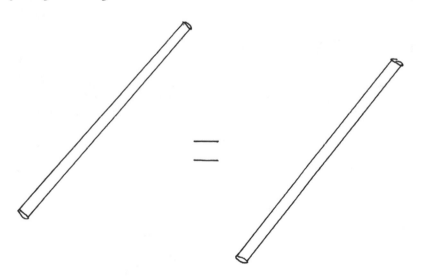

Problem 3.20

The smooth 180-lb pipe has a length of 20 ft and a negligible diameter. It is carried on a truck as shown. Draw the free-body and kinetic diagrams of the pipe.

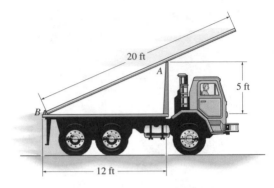

Solution

1. Imagine the pipe to be separated or detached from the system.
2. The pipe is subjected to four *external* forces. They are caused by:

   i. It's weight   **ii. Reaction at** *A*

   iii. <u>Two</u> reactions at *B*

3. Draw the free-body diagram of the (detached) pipe showing all these forces labeled with their magnitudes and directions. Include any other relevant information e.g. lengths, angles etc. which may help when formulating the equations of motion (including the moment equation) for the pipe.
4. Draw the corresponding kinetic diagram.

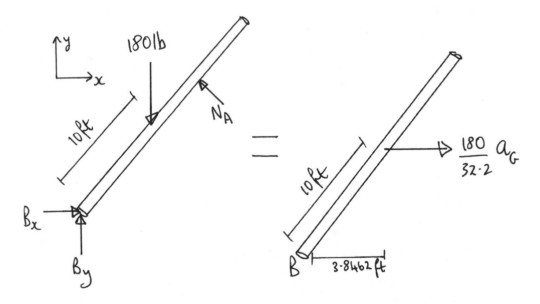

Problem 3.21

The van has a weight of 4500 lb and a center of gravity at G_v. It carries a fixed 800 lb load which has center of gravity at G_l. The van is travelling at 40 ft/s when the brakes are applied causing all the wheels to lock or skid. The coefficient of kinetic friction betwen the wheels and the pavement is $\mu_k = 0.3$. Draw the free-body and kinetic diagrams of the van. Neglect the mass of the wheels. Use these diagrams to write down equations of motion for the van and hence calculate the deceleration of the van as it skids to a complete stop.

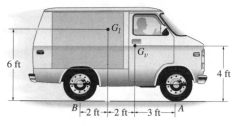

Solution

1. Imagine the van to be separated or detached from the system.
2. The van (including load) is subjected to six *external* forces. They are caused by:

   i.   ii.

   iii.   iv.

   v.   vi.

3. Draw the free-body diagram of the (detached) van showing all these forces labeled with their magnitudes and directions. Include any other relevant information e.g. lengths, angles etc. which may help when formulating the equations of motion (including the moment equation) for the van.
4. Draw the corresponding kinetic diagram.

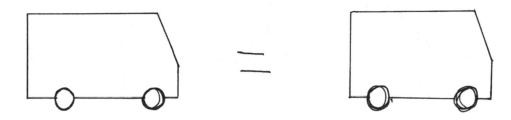

5. Using the inertial coordinate system chosen on the free-body diagram write down the equations of motion:

$$\xleftarrow{} + \sum F_x = m(a_G)_x:$$
$$+ \uparrow \sum F_y = m(a_G)_y:$$

6. Solve for the acceleration of the van:

Problem 3.21.

The van has a weight of 4500 lb and a center of gravity at G_v It carries a fixed 800 lb load which has center of gravity at G_l. The van is travelling at 40 ft/s when the brakes are applied causing all the wheels to lock or skid. The coefficient of kinetic friction betwen the wheels and the pavement is $\mu_k = 0.3$. Draw the free-body and kinetic diagrams of the van. Neglect the mass of the wheels. Use these diagrams to write down equations of motion for the van and hence calculate the deceleration of the van as it skids to a complete stop.

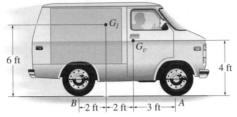

Solution

1. Imagine the van to be separated or detached from the system.

2. The van (including load) is subjected to six *external* forces. They are caused by:

   i. **It's weight**   ii. **Reaction at** A

   iii. **Reaction at** B iv. **Friction at** A

   v. **Friction at** B vi. **Weight of load** $\mathbf{W}_L$

3. Draw the free-body diagram of the (detached) van showing all these forces labeled with their magnitudes and directions. Include any other relevant information e.g. lengths, angles etc. which may help when formulating the equations of motion (including the moment equation) for the van.

4. Draw the corresponding kinetic diagram.

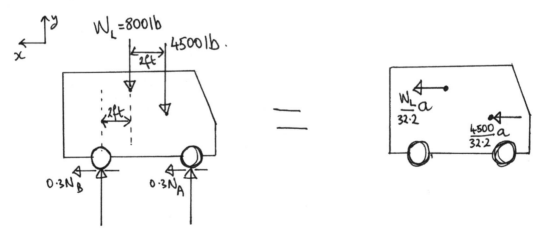

5. Using the inertial coordinate system chosen on the free-body diagram write down the equations of motion:

$$\leftarrow + \sum F_x = m(a_G)_x: \quad 0.3N_B + 0.3N_A = \frac{W_L}{32.2}a + \frac{4500}{32.2}a$$

$$+ \uparrow \sum F_y = m(a_G)_y: \quad N_B + N_A - W_L - 4500 = 0$$

6. Solve for the acceleration of the van:

$$\text{Set } W_L = 800 \text{ lb and obtain } \boldsymbol{a} = 9.66 \text{ ft/s}^2 \leftarrow \qquad \textbf{Ans.}$$

Problem 3.22

The forks on the tractor support the pallet that carries a mass of 400 kg. The load, which is initially at rest, is subjected to curvilinear translation with a radius of 3 m as it is lowered with the maximum initial angular acceleration permitted to prevent it from slipping. The coefficient of static friction between the pallet and the forks is $\mu_s = 0.4$. Draw free-body and kinetic diagrams for the load. Use these diagrams to write down equations of motion for the load and hence calculate the required initial maximum angular acceleration.

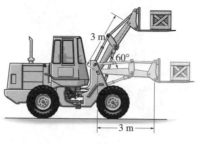

Solution

1. Imagine the load to be separated or detached from the system.
2. The load is subjected to three *external* forces. They are caused by:

   i,   ii.

   iii.

3. Draw the free-body diagram of the (detached) load showing all these forces labeled with their magnitudes and directions. Include any other relevant information e.g. lengths, angles etc. which may help when formulating the equations of motion (including the moment equation) for the load.
4. Draw the corresponding kinetic diagram.

5. Using the inertial coordinate system chosen on the free-body diagram write down the equations of motion:

$$\rightarrow + \sum F_x = m(a_G)_x:$$
$$+ \downarrow \sum F_y = m(a_G)_y:$$

6. Solve for the angular acceleration of the load:

Problem 3.22

The forks on the tractor support the pallet that carries a mass of 400 kg. The load, which is initially at rest, is subjected to curvilinear translation with a radius of 3 m as it is lowered with the maximum initial angular acceleration permitted to prevent it from slipping. The coefficient of static friction between the pallet and the forks is $\mu_s = 0.4$. Draw free-body and kinetic diagrams for the load. Use these diagrams to write down equations of motion for the load and hence calculate the required initial maximum angular acceleration.

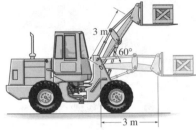

Solution

1. Imagine the load to be separated or detached from the system.
2. The load is subjected to three *external* forces. They are caused by:

    i. **It's weight**   ii. **Reaction from surface**

    iii. **Friction at surface**

3. Draw the free-body diagram of the (detached) load showing all these forces labeled with their magnitudes and directions. Include any other relevant information e.g. lengths, angles etc. which may help when formulating the equations of motion (including the moment equation) for the load.
4. Draw the corresponding kinetic diagram.

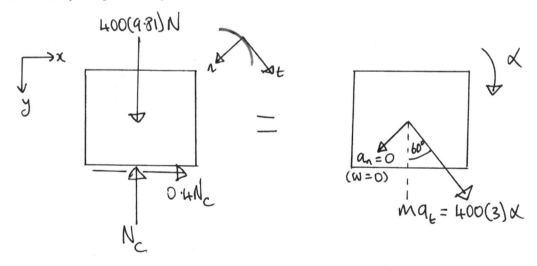

5. Using the inertial coordinate system chosen on the free-body diagram write down the equations of motion:

$$\rightarrow + \sum F_x = m(a_G)_x: \quad 0.4N_C = 400(3)\alpha \sin 60°$$

$$+ \downarrow \sum F_y = m(a_G)_y: \quad -N_C + 400(9.81) = 400(3)\cos 60°$$

6. Solve for the angular acceleration of the load:

$$N_C = 3.19 kN, \quad \alpha = 1.23 \text{ rad/s}^2 \curvearrowright \qquad\qquad\qquad \textbf{Ans.}$$

Problem 3.23

The uniform bar BC has a weight of 40 lb and is pin-connected to the two links which have negligible mass. Draw free-body and kinetic diagrams for the bar BC at the instant $\theta = 30°$.

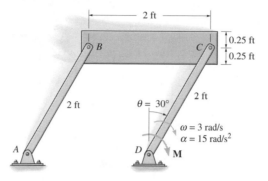

Solution

1. Imagine the bar to be separated or detached from the system.
2. The bar is subjected to five *external* forces. They are caused by:

    i.   **ii.**

    iii.   **iv.**

    v.

3. Draw the free-body diagram of the (detached) bar showing all these forces labeled with their magnitudes and directions. Include any other relevant information e.g. lengths, angles etc. which may help when formulating the equations of motion (including the moment equation) for the pipe.
4. Draw the corresponding kinetic diagram.

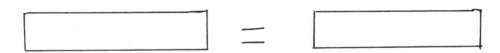

Problem 3.23

The uniform bar BC has a weight of 40 lb and is pin-connected to the two links which have negligible mass. Draw free-body and kinetic diagrams for the bar BC at the instant $\theta = 30°$.

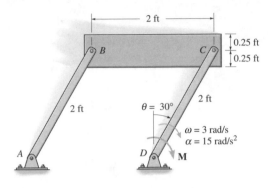

Solution

1. Imagine the bar to be separated or detached from the system.
2. The bar is subjected to five *external* forces. They are caused by:

   i. **It's weight**   ii. **Two reactions at** C

   iii. **Two reactions at** B

3. Draw the free-body diagram of the (detached) bar showing all these forces labeled with their magnitudes and directions. Include any other relevant information e.g. lengths, angles etc. which may help when formulating the equations of motion (including the moment equation) for the pipe.
4. Draw the corresponding kinetic diagram.

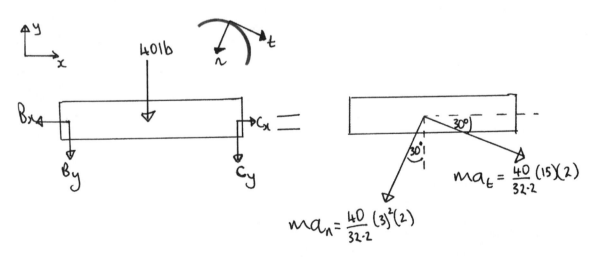

Problem 3.24

The 80 kg disk is supported by a pin at A and is rotating clockwise at $\omega = 0.5$ rad/s when it is in the position shown. Draw the free-body and kinetic diagrams of the disk at this instant.

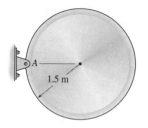

Solution

1. Imagine the disk to be separated or detached from the system.
2. The disk is subjected to three *external* forces. They are caused by:

   i.   **ii.**

   iii.

3. Draw the free-body diagram of the (detached) disk showing all these forces labeled with their magnitudes and directions. Include any other relevant information e.g. lengths, angles etc. which may help when formulating the equations of motion (including the moment equation) for the disk.
4. Draw the corresponding kinetic diagram indicating clearly the acceleration components of the disk.

Problem 3.24

The 80 kg disk is supported by a pin at A and is rotating clockwise at $\omega = 0.5$ rad/s when it is in the position shown. Draw the free-body and kinetic diagrams of the disk at this instant.

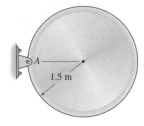

1.5 m

Solution

1. Imagine the disk to be separated or detached from the system.
2. The disk is subjected to three *external* forces. They are caused by:

   i. It's weight   **ii. Two reactions at A**

3. Draw the free-body diagram of the (detached) disk showing all these forces labeled with their magnitudes and directions. Include any other relevant information e.g. lengths, angles etc. which may help when formulating the equations of motion (including the moment equation) for the disk.
4. Draw the corresponding kinetic diagram indicating clearly the acceleration components of the disk.

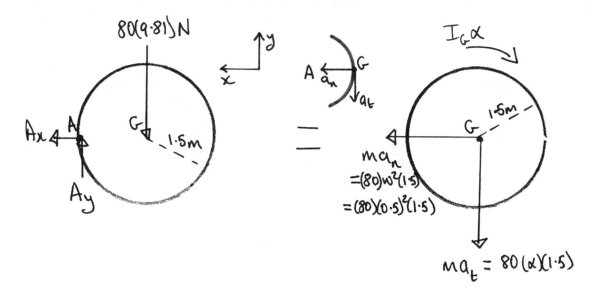

Problem 3.25

The drum has a weight of 20 lb and a radius of gyration about its mass center of 0.8 ft. If the block has a weight of 12 lb, draw the free-body and kinetic diagrams of the drum and block system and use them to determine the angular acceleration of the drum if the block is allowed to fall freely.

Solution

1. Imagine the drum and block system to be separated or detached from the pin at A.
2. The drum and block system is subjected to four *external* forces. They are caused by:

   i.   II.

   iii.   iv.

3. Draw the free-body diagram of the (detached) drum and block system showing all these forces labeled with their magnitudes and directions. Include any other relevant information e.g. lengths, angles etc. which may help when formulating the equations of motion (including the moment equation) for the drum and block system.
4. Draw the corresponding kinetic diagram indicating clearly the acceleration components of the drum and block.

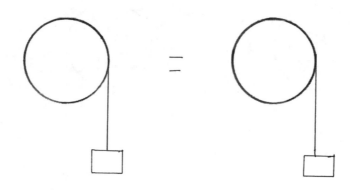

5. Sum moments about A (why?) and write down the moment equation of motion:

   $$\circlearrowleft + \sum M_A = \sum (M_k)_A:$$

6. Solve for the angular acceleration α_D of the drum:

Problem 3.25

The drum has a weight of 20 lb and a radius of gyration about its mass center of 0.8 ft. If the block has a weight of 12 lb, draw the free-body and kinetic diagrams of the drum and block system and use them to determine the angular acceleration of the drum if the block is allowed to fall freely.

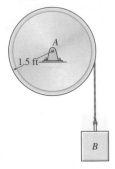

Solution

1. Imagine the drum and block system to be separated or detached from the pin at A.

2. The drum and block system is subjected to four *external* forces. They are caused by:

    i. Weight of drum   **ii. Two reactions at A**

    iii. Weight of block

3. Draw the free-body diagram of the (detached) drum and block system showing all these forces labeled with their magnitudes and directions. Include any other relevant information e.g. lengths, angles etc. which may help when formulating the equations of motion (including the moment equation) for the drum and block system.

4. Draw the corresponding kinetic diagram indicating clearly the acceleration components of the drum and block.

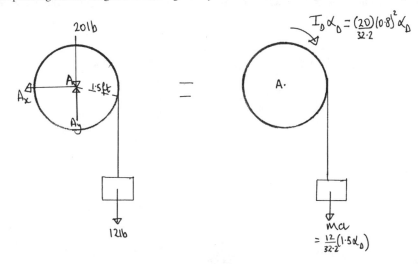

5. Sum moments about A (why? — **eliminates pin reactions**) and write down the moment equation of motion:

$$\circlearrowleft + \sum M_A = \sum (M_k)_A: \quad 12(1.5) = \left[\frac{20}{32.2}(0.8)^2\right]\alpha_D + \left[\frac{12}{32.2}(1.5\alpha_D)\right](1.5)$$

6. Solve for the angular acceleration α_D of the drum:

$$\alpha_D = 14.6 \text{ rad/s}^2 \ \curvearrowright \qquad\qquad\qquad\qquad\qquad\qquad \textbf{Ans.}$$

Problem 3.26

The 10 lb bar is pinned at its center O and connected to a torsional spring. The spring has a stiffness $k = 5$ lb.ft/rad so that the torque developed is $M = (5\theta)$ lb.ft, where θ is in radians. The bar is released from rest when $\theta = 90°$. Draw a free-body diagram for the bar when $\theta = 45°$.

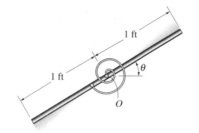

Solution

1. Imagine the bar to be separated or detached from the system.
2. The bar is subjected to three *external* forces and one applied couple moment. They are caused by:

   i.   **ii.**

   iii.   **iv.**

3. Draw the free-body diagram of the (detached) bar showing all these forces labeled with their magnitudes and directions. Include any other relevant information e.g. lengths, angles etc. which may help when formulating the equations of motion (including the moment equation) for the bar.
4. Indicate the acceleration components of the bar on the coordinate axes system chosen in the free-body diagram.

Problem 3.26

The 10 lb bar is pinned at its center O and connected to a torsional spring. The spring has a stiffness $k = 5$ lb.ft/rad so that the torque developed is $M = (5\theta)$ lb.ft, where θ is in radians. The bar is released from rest when $\theta = 90°$. Draw a free-body diagram for the bar when $\theta = 45°$.

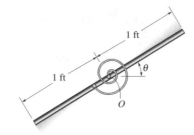

Solution

1. Imagine the bar to be separated or detached from the system.
2. The bar is subjected to three *external* forces and one applied couple moment. They are caused by:

   i. It's weight   **ii. Two reactions at O**

   iii. Applied torque M

3. Draw the free-body diagram of the (detached) bar showing all these forces labeled with their magnitudes and directions. Include any other relevant information e.g. lengths, angles etc. which may help when formulating the equations of motion (including the moment equation) for the bar.
4. Indicate the acceleration components of the bar on the coordinate axes system chosen in the free-body diagram.

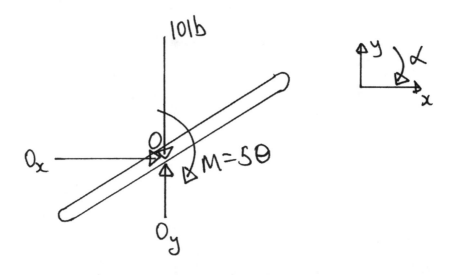

Problem 3.27

The 20 kg roll of paper has a radius of gyration $k_A = 90$ mm about an axis passing through point A. It is pin-supported at both ends by two brackets AB. The roll rests against a wall for which the coefficient of kinetic friction is $\mu_k = 0.2$. A constant vertical force of magnitude F is applied to the roll to pull off 1 m of paper starting from rest. Draw a free-body diagram for the roll of paper. Neglect the mass of paper that is removed.

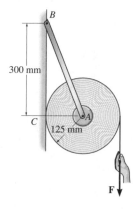

Solution

1. Imagine the roll to be separated or detached from the system.
2. The bar is subjected to five *external* forces (use one force to descibe the effect of AB on the roll) They are caused by:

   i.   ii.

   iii.   iv.

   v.

3. Draw the free-body diagram of the (detached) roll showing all these forces labeled with their magnitudes and directions. Include any other relevant information e.g. lengths, angles etc. which may help when formulating the equations of motion (including the moment equation) for the roll.

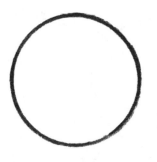

Problem 3.27

The 20 kg roll of paper has a radius of gyration $k_A = 90$ mm about an axis passing through point A. It is pin-supported at both ends by two brackets AB. The roll rests against a wall for which the coefficient of kinetic friction is $\mu_k = 0.2$. A constant vertical force of magnitude F is applied to the roll to pull off 1 m of paper starting from rest. Draw a free-body diagram for the roll of paper. Neglect the mass of paper that is removed.

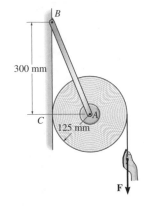

Solution

1. Imagine the roll to be separated or detached from the system.

2. The bar is subjected to five *external* forces (use one force to descibe the effect of AB on the roll) They are caused by:

   i. **It's weight**   ii. **Reaction at C**

   iii. **Friction at C**   iv. **Force in bar AB**

   v. **Applied force F**

3. Draw the free-body diagram of the (detached) roll showing all these forces labeled with their magnitudes and directions. Include any other relevant information e.g. lengths, angles etc. which may help when formulating the equations of motion (including the moment equation) for the roll.

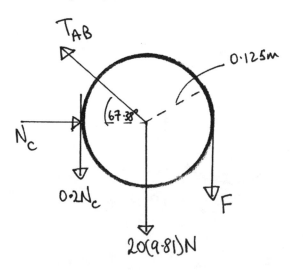

Problem 3.28

The door will close automatically using torsional springs mounted on the hinges. Each spring has a stiffness $k = 50$ N.m/rad so that the torque on each hinge is $M = (50\theta)$ N.m where θ is measured in radians. The door is released from rest when it is open at $\theta = 90°$. Treating the door as a thin plate having a mass of 70 kg, draw a free-body diagram for the door at the instant $\theta = 0°$.

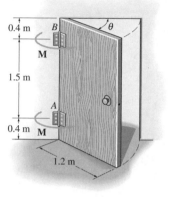

Solution

1. Imagine the door to be separated or detached from the system.
2. The door is subjected to five *external* forces and an applied couple moment. They are caused by:

   i.   ii.

   iii.   iv.

   v.   vi.

3. Draw the free-body diagram of the (detached) door showing all these forces labeled with their magnitudes and directions. Include any other relevant information e.g. lengths, angles etc. which may help when formulating the equations of motion (including the moment equation) for the door.

Problem 3.28

The door will close automatically using torsional springs mounted on the hinges. Each spring has a stiffness $k = 50$ N.m/rad so that the torque on each hinge is $M = (50\theta)$ N.m where θ is measured in radians. The door is released from rest when it is open at $\theta = 90°$. Treating the door as a thin plate having a mass of 70 kg, draw a free-body diagram for the door at the instant $\theta = 0°$.

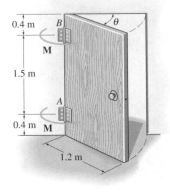

Solution

1. Imagine the door to be separated or detached from the system.
2. The door is subjected to five *external* forces and an applied couple moment. They are caused by:

   i. **It's weight**   ii. **Two reactions at** A

   iii. **Two reactions at** B iv. **Applied torque**

3. Draw the free-body diagram of the (detached) door showing all these forces labeled with their magnitudes and directions. Include any other relevant information e.g. lengths, angles etc. which may help when formulating the equations of motion (including the moment equation) for the door.

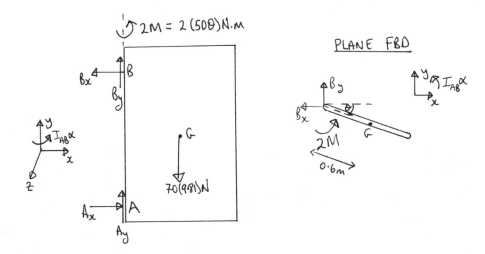

Problem 3.29

Cable is unwound from a spool supported on small rollers at A and B by exerting a force $T = 300$ N on the cable. The spool and cable have a total mass of 600 kg and a radius of gyration of $k_o = 1.2$ m. Draw a free-body diagram for the spool and use it to compute the time needed to unravel 5 m of cable from the spool. Neglect the mass of the cable being unwound and the mass of the rollers at A and B. The rollers turn with no friction.

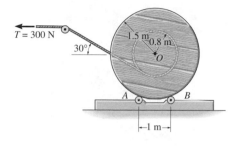

Solution

1. Imagine the spool to be separated or detached from the system.
2. The spool is subjected to four *external* forces. They are caused by:

   i.   ii.

   iii.   iv.

3. Draw the free-body diagram of the (detached) spool showing all these forces labeled with their magnitudes and directions. Include any other relevant information e.g. lengths, angles etc. which may help when formulating the equations of motion (including the moment equation) for the spool.

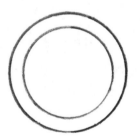

4. Sum moments about the center of the spool (O) and write down the moment equation of motion:

$$\circlearrowleft + \sum M_O = I_O \alpha:$$

5. Solve for the angular acceleration α of the spool.
6. Use kinematics (in terms of angular displacement) to solve for the required time t.

Problem 3.29

Cable is unwound from a spool supported on small rollers at A and B by exerting a force $T = 300$ N on the cable. The spool and cable have a total mass of 600 kg and a radius of gyration of $k_o = 1.2$ m. Draw a free-body diagram for the spool and use it to compute the time needed to unravel 5 m of cable from the spool. Neglect the mass of the cable being unwound and the mass of the rollers at A and B. The rollers turn with no friction.

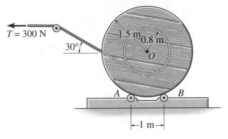

Solution

1. Imagine the spool to be separated or detached from the system.
2. The spool is subjected to four *external* forces. They are caused by:

      i. **It's weight**   ii. **Reaction at A**

      iii. **Reaction at B**   iv. **Applied force T**

3. Draw the free-body diagram of the (detached) spool showing all these forces labeled with their magnitudes and directions. Include any other relevant information e.g. lengths, angles etc. which may help when formulating the equations of motion (including the moment equation) for the spool.

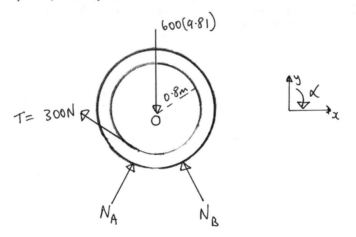

4. Sum moments about the center of the spool (O) and write down the moment equation of motion:

$$\curvearrowleft + \sum M_O = I_O\alpha: \quad 300(0.8) = 600(1.2)^2\alpha$$

5. Solve for the angular acceleration α of the spool: $\alpha = 0.2778 rad/s^2 \curvearrowright$.
6. Use kinematics (in terms of angular displacement θ) to solve for the required time t.

$$\theta = \theta_0 + \omega_0 t + \frac{1}{2}\alpha t^2. \quad \text{Set } \theta = \frac{s}{r} = \frac{5}{0.8} = 6.25 \text{ rad.}$$

$$6.25 = 0 + 0 + \frac{1}{2}(0.2778)t^2 \Rightarrow t = 6.71 \text{ s} \qquad\qquad \textbf{Ans.}$$

Problem 3.30

The disk has a mass M and a radius R. If a block of mass m is attached to the cord, draw free-body and kinetic diagrams of the disk and mass system and use them to determine the angular acceleration of the disk when the block is released from rest.

Solution

1. Imagine the disk and block system to be separated or detached from the pin at the center of the disk.
2. The disk and block system is subjected to four *external* forces. They are caused by:

   i.   **ii.**

   iii.   **iv.**

3. Draw the free-body diagram of the (detached) disk and block system showing all these forces labeled with their magnitudes and directions. Include any other relevant information e.g. lengths, angles etc. which may help when formulating the equations of motion (including the moment equation) for the disk and block system.
4. Draw the corresponding kinetic diagram indicating clearly the acceleration components of the disk and block.

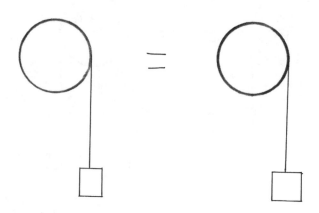

5. Sum moments about the center of the disk (O) (why?) and write down the moment equation of motion:

$$\circlearrowleft \, + \sum M_O = \sum (M_k)_O:$$

6. Solve for the angular acceleration α_D of the disk:

Problem 3.30

The disk has a mass M and a radius R. If a block of mass m is attached to the cord, draw free-body and kinetic diagrams of the disk and mass system and use them to determine the angular acceleration of the disk when the block is released from rest.

Solution

1. Imagine the disk and block system to be separated or detached from the pin at the center of the disk.
2. The disk and block system is subjected to four *external* forces. They are caused by:

   i. Weight of disk   **ii. Two reactions at A**

   iii. Weight of block

3. Draw the free-body diagram of the (detached) disk and block system showing all these forces labeled with their magnitudes and directions. Include any other relevant information e.g. lengths, angles etc. which may help when formulating the equations of motion (including the moment equation) for the disk and block system.
4. Draw the corresponding kinetic diagram indicating clearly the acceleration components of the disk and block.

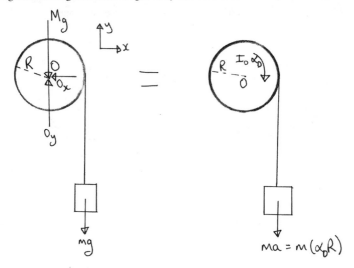

5. Sum moments about the center of the disk (O) (why? — **eliminates pin reactions**) and write down the moment equation of motion:

$$\circlearrowleft + \sum M_O = \sum (M_k)_O: \quad mgR = \frac{1}{2}MR^2(\alpha_D) + m(\alpha_D R)R$$

6. Solve for the angular acceleration α_D of the disk:

$$\alpha_D = \frac{2mg}{R(M + 2m)} \quad \circlearrowleft \qquad\qquad\qquad \textbf{Ans.}$$